# CliffsQuickReview® Precalculus

**By W. Michael Kelley**

***About the Author***

Mike Kelley has been a high school and college math instructor. He currently works as an Academic Technology Coordinator for the College of Education at the University of Maryland. He has written several books and owns the Web site www.calculus-help.com.



***Publisher's Acknowledgments***

***Editorial***

Senior Acquisitions Editor: Greg Tubach
Project Editor: Tim Ryan
Development Editor: Ted Lorenzen
Copy Editor: Elizabeth Welch
Technical Editor: Jeff Poet, PhD
Editorial Assistant: Amanda Harbin

***Composition***

Indexer: Tom Dinse
Proofreader: Ethel M. Winslow
Wiley Indianapolis Composition Services

**CliffsQuickReview® *Precalculus***

Published by:

Learneo, Inc.
2000 Seaport Blvd, 3rd Floor
Redwood City, CA 94063
United States of America





Published simultaneously in Canada

ISBN: 978-1-957671-19-2

Printed in the United States of America

10 9

1O/RV/RQ/QY/IN

# Table of Contents

# INTRODUCTION

*CliffsQuickReview Precalculus* is a comprehensive volume of the topics usually included within a course intended to serve as a calculus prerequisite. Although the collection of skills deemed worthy of inclusion in such a course may vary slightly from instructor to instructor, this text contains all of the most commonly discussed elements, including:

- Arithmetic and algebraic skills
- Functions and their graphs
- Polynomials, including binomial expansion
- Right and oblique angle trigonometry
- Equations and graphs of conic sections
- Matrices and their application to systems of equations

It is assumed that you have some knowledge of algebra and its concepts, although nearly all of the foundational algebraic skills you'll need are reviewed in the early chapters of this book. If you feel you need to further review these concepts, refer to *CliffsQuickReview Algebra I* and *Algebra II.*

## Why You Need This Book

Can you answer yes to any of these questions?

- Do you need to review the fundamentals of precalculus?
- Do you wish you had someone else to explain the concepts of precalculus to you other than your teacher?
- Do you need to prepare for a precalculus test?
- Do you need a concise, comprehensive reference for precalculus?

If so, then *CliffsQuickReview Precalculus* is for you!

## How to Use This Book

This book puts you in the driver's seat; use it any way you like. Perhaps you want to read about the topics you'll learn in class before your teacher discusses them, so that you have a leg up on your classmates and stand a better chance to understand since you'll already have an idea about what's going on. Maybe you want to read the book cover to cover, or just consult it when you're having trouble understanding what's going on in class. Either way, here are a few ways you can search for more information about a particular topic:

- Use the Pocket Guide (the tear-out card in the front of the book) to find essential information, such as key formulas and concepts.
- Look for areas of interest in the Table of Contents, or use the index to find specific topics.
- Flip through the book, looking for subject areas at the top of each page.
- Get a glimpse of what you'll gain from a chapter by reading through the "Chapter Check-In" at the beginning of each chapter.
- Use the Chapter Checkout at the end each chapter to gauge your grasp of the important information you need to know.
- At the end of the book, look for additional sources of information in the CQR Resource Center.
- Use the glossary to find key words quickly. Terms are written in **boldface** when first introduced in the book, so their definitions are always close by. In addition, all of the important boldface terms are defined in the book's glossary.

## Visit Our Web Site

Make sure to look us up on the Web at www.cliffsnotes.com; we host an extremely valuable site featuring review materials, top-notch Internet links, quizzes, and more to enhance your learning. The site also features timely articles and tips, plus downloadable versions of any CliffsNotes books.

When you stop by, don't hesitate to share your thoughts about this book or any John Wiley & Sons product. Just click the "Talk to Us" button. We welcome your feedback!

# Chapter 1
# PRECALCULUS PREREQUISITES

## Chapter Check-In

- ❑ Defining common number groups
- ❑ Writing inequalities as intervals
- ❑ Understanding algebraic properties
- ❑ Working with exponents, radicals, polynomials, and rational expressions
- ❑ Finding solutions to equations and inequalities
- ❑ Constructing linear equations

A strong algebraic background is essential to success in precalculus. Before you can begin exploring its more advanced topics, you must first have a firm grip on the fundamentals. In this chapter, you'll review and practice foundational concepts and skills.

## Classifying Numbers

Many times throughout your precalculus course, you'll be manipulating specific kinds of numbers, so it's important to understand how mathematicians classify numbers and what kinds of major classifications exist. Be aware that numbers can fall into more than one group. Just as an American citizen can also be classified as a North American citizen or a citizen of Earth, numbers may also belong to numerous categories simultaneously.

The following groups, or sets, of numbers are generally agreed on by mathematicians as the most common classifications of numbers. They are listed here in order of size, from smallest to largest:

- **Natural numbers.** The set of numbers you've used since you were very young when counting (as such, the *natural numbers* can also be called the *counting numbers*): {1, 2, 3, 4, 5, 6, ...}.

- **Whole numbers.** The *whole numbers* include all of the *natural numbers* and, also, the number 0: {0, 1, 2, 3, 4, 5, ...}.
- **Integers.** All of the whole numbers and their opposites make up the set of *integers.* In other words, any number without an extra decimal or fraction attached to it is considered an integer: {..., –3, –2, –1, 0, 1, 2, 3, ...}. Because *integers* contain no obvious fractions or decimals, some students are tempted to refer to them as *whole numbers,* but that is not completely accurate, because the set of *whole numbers* does not include negative numbers.
- **Rational numbers.** A number is classified as *rational* if one of the following conditions hold true.

  The number can be expressed as a fraction. (In other words, the number can be rewritten as $\frac{a}{b}$, where $a$ and $b$ are integers, and $b \neq 0$.)

  The number is a terminating decimal, in other words a decimal that ends (such as 6.25) rather than continues on infinitely.

  The number is a decimal that repeats in an infinite pattern (such as 5.297297297297...).

  Basically, any number that can be written as a fraction is *rational.*

**Example 1:** Show that any integer must also be a rational number.

Any integer $a$ can also be rewritten as $\frac{a}{1}$, since dividing by 1 will not alter the value of the integer. Because $a$ can be expressed as a fraction whose numerator and denominator are both integers, $a$ must be rational by definition.

- **Irrational numbers.** A number that cannot be written as a fraction is considered *irrational.* The most obvious indicator of an irrational number is a decimal that doesn't infinitely repeat itself yet never terminates. For example, $\pi$ is an irrational number whose decimal equivalent 3.14159265359... never ends and never follows any obvious repeating pattern. Many radicals, like $\sqrt{3}$, are irrational numbers.
- **Real numbers.** The *real numbers* are made up of the *rational numbers* and the *irrational numbers* grouped together, as shown in Figure 1-1.
- **Complex numbers.** *Complex numbers* differ from the *real numbers* in appearance quite starkly. Complex numbers usually have two distinct parts and look like $a + bi$, where $a$ is the *real part*, $bi$ is the *imaginary part*, where $i$ is equal to the imaginary value $\sqrt{-1}$. However,

numbers need not contain both parts to be considered complex. In fact, any real number is automatically complex. For example, since the real number 3 can be written as $3 + 0i$, 3 is a complex number. It just contains no imaginary part.

**Figure 1-1** The rational and the irrational numbers together comprise the entire set of real numbers. Note that this drawing is not to scale. Far more irrational than rational numbers exist.

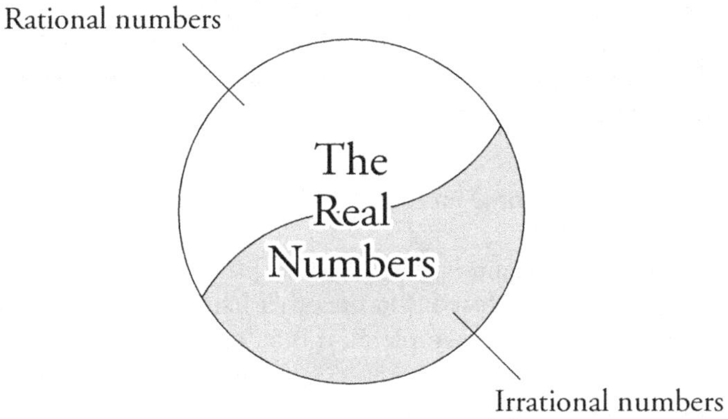


## Interval Notation

Traditional inequality statements can be rewritten using **interval notation**, a shorthand method that expresses the same meaning but usually in a more compact and intuitive manner. This is largely due to the fact that interval notation clearly defines the boundaries of the inequality with which you're working.

### Bounded intervals

If you're given an inequality that is bounded on both sides by a real number, that statement can be rewritten as a **bounded interval**. To create a bounded interval, write the two numerical endpoints of the interval in order, always from lowest to highest. (The interval will almost look like a coordinate pair.) Then, indicate whether that point should be included on the interval. If it should, use a bracket with that endpoint; if it should not, use a parenthesis.

**Example 2:** Rewrite the following inequality statements using interval notation.

(a) $-5 \le x \le 3$

Because the inequality signs stipulate less than *or equal to*, you must include the endpoints in the interval. Had equality not been a possibility, you would not include those endpoints. Use brackets to indicate inclusion: [–5,3].

(b) $1 > x > 0$

Even though this interval is written so that the upper boundary is on the left, interval notation still requires you to write them in order from lesser to greater. Use parentheses to indicate that the endpoints are not included in the interval: (0,1).

(c) $-2 \le x < 4$

The lower endpoint is included while the upper is not: [–2,4).

If both endpoints of the interval are included (as in part [a] of Example 2), the interval is said to be **closed**. On the other hand, if neither endpoint is included (as in part [b] of Example 3), it is an **open interval**.

## Unbounded intervals

Sometimes, only one endpoint of an interval is explicitly defined and the other is implied. For instance, consider the inequality $x > 3$. Clearly, the lower boundary of the interval is 3, but what is the upper boundary? Because there is no finite value given for the upper endpoint, you use infinity. If one or more of the endpoints of an interval are understood to be infinite, the interval is said to be **unbounded**.

You will use two different infinite boundaries:

- $\infty$, if the boundary is infinitely positive (it is used as the upper bound of the interval)
- $-\infty$, if the boundary is infinitely negative (it is used as the lower bound of the interval)

Infinity is technically not a real number, which means you can never use a bracket to indicate its inclusion in the interval. Instead, always use a parenthesis.

**Example 3:** Rewrite the following inequality statements using interval notation:

(a) $x > -1$

The lower bound of the interval is –1, and the upper bound is infinitely large, since any positive number will make this inequality statement true. The lower boundary should not be included, because the relationship is "greater than," not "greater than or equal to": $(-1,\infty)$.

(b) $x \leq 3$

In this interval, 3 is the upper boundary. If the lower boundary of an interval is infinite, you must indicate this by using negative infinity: $(-\infty,3]$.

(c) All real numbers

Any real number, from the infinitely negative to the infinitely positive, should be included in this interval: $(-\infty,\infty)$.

# Algebraic Properties

**Properties**, also called *laws* or *axioms*, are foundational mathematical principles that are assumed true. Although there is no way to irrefutably prove properties, they make enough inherent common sense to be universally agreed on by mathematicians. It's a good thing they are, because these laws form the backbone of algebra.

## The associative property

Given a string of numbers added together, you may group the numbers in any order you wish and it will not affect the answer you get. This is the basic premise of the **associative property** for addition. In other words, no matter what numbers are *associated* together, you will get the same result in the end.

$$(1 + 3) + 5 = 1 + (3 + 5)$$
$$4 + 5 = 1 + 8$$
$$9 = 9$$

The associative property also holds true for multiplication, but it fails for both subtraction and division. Here are the official mathematical definitions for its two incarnations:

- **The associative property for addition:**

  $(a + b) + c = a + (b + c)$

- **The associative property for multiplication:**

  $(a \cdot b) \cdot c = a \cdot (b \cdot c)$

Note that the symbol · is used here to indicate multiplication rather than the other traditional symbol for multiplication, ×. This is because it's easy to confuse the operation × with the variable *x* when you're working a problem.

### The commutative property

This property (like its sister, the associative property) works only for addition and multiplication. In essence, it says that given a string of numbers being added or a string of numbers being multiplied, the order in which you complete that operation doesn't matter.

$$3 \cdot 9 = 9 \cdot 3$$
$$27 = 27$$

- **The commutative property for addition:**

  $a + b = b + a$

- **The commutative property for multiplication:**

  $a \cdot b = b \cdot a$

### The distributive property

According to the **distributive property**, if terms are being added or subtracted within parentheses and a number appears "outside" that group of terms, you can multiply that outer number through to every number within those parentheses.

$$a(b + c) = ab + ac$$

**Example 4:** Rewrite using the distributive property:

$$3(x - 7)$$

Multiply every term in the parentheses by 3:

$$3 \cdot x - 3 \cdot 7$$
$$3x - 21$$

### Identity elements

Fixed numbers called **identity elements** exist for both the operations of addition and multiplication. These elements do not alter a number's value (or *identity*) when the operation is applied to them. The *identity element for addition* (also called the *additive identity*) is 0, because if you add 0 to any number, you get back what you started with:

$$2 + 0 = 2$$

Similarly, the *identity element for multiplication* (also called the *multiplicative identity*) is 1, since multiplying any number by 1 will not change that number's value:

$$3 \cdot 1 = 3$$

These identity elements are important because they are a major component in the *inverse properties.*

### Inverse properties

Once again, the operations of addition and multiplication have a property specific to them. In both cases, an **inverse property** assures you that no matter the input, there is a way to "cancel it out."

- **Additive inverse property:** For any real number $a$, there exists a real number $-a$ (the *opposite* of $a$) so that $a + (-a) = 0$:

$$4 + (-4) = 0$$

- **Multiplicative inverse property:** For any non-zero real number $a$, there exists a real number $\frac{1}{a}$ so that $a \cdot \frac{1}{a} = 1$:

$$7 \cdot \frac{1}{7} = 1$$

Note that when you "undo" addition and multiplication using these inverse properties, the result will be the *identity element* for the corresponding operation.

## Exponential Expressions

Repeated multiplication can be rewritten using **exponents**, small numbers written above and to the right of the **base** number, both to clarify and simplify your notation. Rather than write "$x \cdot x \cdot x$," you can write "$x^3$," which is read "$x$ to the third **power**." The *power* of an exponent is the number of times the object is multiplied by itself. Therefore, in the expression $x^3$, $x$ is considered the *base* and 3 is the *power.*

There are six important rules you should know when undertaking any arithmetic involving exponents:

- **Rule 1:** $x^a \cdot x^b = x^{a+b}$

  If two exponential expressions with identical bases are multiplied, the result is that base raised to an exponent equal to the sum of the two powers:

  $$x^4 \cdot x^7 = x^{4+7} = x^{11}$$

- **Rule 2:** $\frac{x^a}{x^b} = x^{a-b}$

  If two exponential expressions with identical bases are divided, the result is that base raised to an exponent equal to the power in the numerator minus the power in the denominator:

  $$\frac{x^8}{x^5} = x^{8-5} = x^3$$

- **Rule 3:** $\left(x^a\right)^b = x^{a \cdot b}$

  If an exponential expression is itself raised to a power, the result is the base raised to the product of the two powers:

  $$\left(x^2\right)^6 = x^{2 \cdot 6} = x^{12}$$

- **Rule 4:** $\left(x^a y^b\right)^c = x^{ac} y^{bc}$

  If numerous exponential factors are raised to a power, multiply the outer power times each of the inner powers.

  $$\left(x^2 y^5\right)^3 = x^{2 \cdot 3} y^{5 \cdot 3} = x^6 y^{15}$$

- **Rule 5:** $x^{-a} = \frac{1}{x^a}$ and $\frac{1}{x^{-b}} = x^b$

  A negative exponent indicates that the expression is in the wrong part of the fraction. To make the exponent positive again (no algebraic expression is completely simplified until it contains no negative exponents), move the exponential expression to the other side of the fraction bar. For instance, if it is in the numerator, move it to the denominator, and leave the base alone.

  $$\frac{x^{-2}}{y^{-3}} = \frac{y^3}{x^2}$$

- **Rule 6:** $x^0 = 1$ (if $x \neq 0$)

  Any real number raised to the 0 power is equal to 1 (with the exception of $0^0$, which does not have a real number value).

**Example 5:** Simplify using the exponential rules:

(a) $\dfrac{x^3y^5z^2}{xy^7z^2}$

Rewrite the fraction using Rule 2. Since the $x$ in the denominator has no visible exponent, it is understood to be 1.

$$x^{3-1}y^{5-7}z^{2-2}$$

$$x^2y^{-2}z^0$$

A completely simplified solution does not contain negative exponents. Apply Rule 5 to achieve that goal. In addition, rewrite $z^0$ as 1.

$$\frac{x^2}{y^2}$$

(b) $(x^2y^3)(x^7yz^3)$

You can rearrange the terms thanks to the commutative propery and then add exponential powers of like bases, thanks to Rule 1. Again, since the $y$ in the second group of parentheses has no exponent visible, it is understood to be 1.

$$x^{2+7}y^{3+1}z^3$$

$$x^9y^4z^3$$

(c) $\left(\dfrac{x^2y^{-2}}{z^{-3}}\right)^{-2}$

Begin by applying Rule 4:

$$\frac{x^{-4}y^4}{z^6}$$

Use Rule 5 to eliminate negative exponents:

$$\frac{y^4}{x^4z^6}$$

# Radical Expressions

Although most of the time the exponents you'll see will be integers, you may run across some fractional powers as well. These types of powers translate into **radicals** (also called *roots*):

$$x^{a/b} = \sqrt[b]{x^a} \text{ or } \left(\sqrt[b]{x}\right)^a$$

You can use either notation to rewrite the fractional power as a radical. In some cases, one form will be more useful than the other when you are simplifying.

A typical radical, $\sqrt[n]{x^a}$ contains two parts: the **index** (the small number in front of the radical) and the **radicand**, the quantity within the radical symbol. It is read "The *n*th root of *x* to the *a*th power." Note that if no *index* is given for the radical, the *index* is understood to be 2.

Some students find radicals easier to understand if they think of the notation as a question. For example, the radical $\sqrt[3]{8}$ asks the question "What number multiplied by itself 3 times is equal to 8?" The answer is 2, so $\sqrt[3]{8} = 2$.

## Properties of radicals

Because radicals are really exponents in disguise (even if they are fractional exponents), radicals possess the same properties as exponents. In addition, radicals have these properties:

- $\sqrt[n]{x^a y^b} = \sqrt[n]{x^a} \cdot \sqrt[n]{y^b}$

  Factors multiplied together inside of a radical can be broken up and written as the product of two radicals with the same index as the original. That is to say, the root of a product is equal to the product of the individual roots:

- $\sqrt[n]{\dfrac{x^a}{y^b}} = \dfrac{\sqrt[n]{x^a}}{\sqrt[n]{y^b}}$

  Just like multiplication, division problems surrounded by radicals can be broken up into separate, smaller radicals as well. So, the root of the quotient is equal to the quotient of the individual roots.

## Simplifying radicals

The most common task you'll face in the study of radicals is the need to simplify radical expressions.

**Example 6:** Use the properties of radicals to simplify these expressions:

(a) $\sqrt{200x^2y}$

Your goal will be to break this radical into two different radicals, one containing all **perfect squares** and the other containing everything else. *Perfect squares* are quantities generated by multiplying some value by itself.

$$\sqrt{100 \cdot 2 \cdot x^2 \cdot y}$$

$$\sqrt{100x^2} \cdot \sqrt{2y}$$

Both 100 and $x^2$ are perfect squares (since $100 = 10 \cdot 10$ and $x^2 = x \cdot x$); the leftmost radical will be eliminated.

$$10|x|\sqrt{2y}$$

You might not have expected the absolute value signs. They are rare but necessary when you have this situation: $\sqrt[n]{x^n}$ and $n$ is an even integer. This precaution ensures that the answer is positive, because a radical with an even index must always be positive.

(b) $\sqrt[3]{-108x^2y^8}$

Again, split up the radical, but this time put all of the **perfect cubes** (values generated by multiplying the same thing by itself three times) together:

$$\sqrt[3]{-27y^6} \cdot \sqrt[3]{4x^2y^2}$$

$$\left(-3y^2\right)\sqrt[3]{4x^2y^2}$$

There is no need to worry about absolute value signs because the index of this radical is odd.

## Operations with radicals

It is a bit more complicated to add and subtract radical expressions than it is to multiply and divide them. In fact, radicals must have the same index and radicand in order to perform addition and subtraction, but that is not the case for multiplication and division.

**Example 7:** Simplify the following expressions:

(a) $5\sqrt{2} - 3\sqrt{8}$

While the indices are the same (they are both 2), the radicands appear different at first glance. That changes when you simplify the expression:

$$5\sqrt{2}-3\sqrt{4}\sqrt{2}$$
$$5\sqrt{2}-6\sqrt{2}$$

Now that they share the same radicand as well, you can combine the coefficients and get $-\sqrt{2}$.

(b) $\left(\sqrt{x}\right)\left(\sqrt[3]{x^2}\right)$

Begin by rewriting the radicals as exponential expressions:

$$x^{1/2}\cdot x^{2/3}$$

Apply Rule 1 for exponential expressions:

$$x^{1/2+2/3}=x^{3/6+4/6}=x^{7/6}=\sqrt[6]{x^7}$$

You may write your final answer in either exponential or radical form; they are equivalent.

### Rationalizing expressions

Some teachers require that you **rationalize** your answers, when appropriate. This means they don't want an answer containing a radical sign in its denominator.

**Example 8:** Rationalize the following fraction:

$$\frac{x}{\sqrt{3}}$$

To eliminate the radical, multiply both the numerator and denominator by a value of $\sqrt{3}$. This is the equivalent of multiplying by 1, so it doesn't change the value of the fraction, and it creates a perfect square in the denominator.

$$\frac{x}{\sqrt{3}}\cdot\frac{\sqrt{3}}{\sqrt{3}}=\frac{x\sqrt{3}}{\sqrt{9}}=\frac{x\sqrt{3}}{3}$$

## Polynomial Expressions

**Polynomials** are strings of **terms** added to or subtracted from one another. Each term is made up of numbers and variables (usually raised to whole number powers) multiplied together. For example, the polynomial

$$4x^3-2x^2+x+7$$

is made up of four terms. The **coefficient** is the numerical value preceding the variable in each term. The first term, $4x^3$, has a *coefficient* of 4, and the second term has a coefficient of –2. The **degree** of this polynomial, defined by the highest exponent found in the polynomial, is 3. The **leading coefficient** is the coefficient accompanying the variable raised to the highest exponential value. In this example, the leading coefficient is 4.

### Classifying polynomials

Polynomials are typically categorized either according to the number of terms they possess or according to the degree of the polynomial.

- **Classifying according to number of terms**

  A polynomial containing only one term is called a **monomial**. If two terms are present, the polynomial is considered a **binomial**, while three terms indicates a **trinomial**. No commonly used terms are available that indicate a polynomial containing four, five, or more terms.

- **Classifying according to degree**

  It's easy to categorize a polynomial according to its degree. Simply look for the highest exponent within the polynomial. Table 1-1 gives the classifications based on a polynomial's degree.

**Table 1-1 Degree Classifications for Polynomials**

| *Degree* | *Category* | *Example* |
|---|---|---|
| 0 | constant | $7$ |
| 1 | linear | $-x + 7$ |
| 2 | quadratic | $5x^2 + x + 7$ |
| 3 | cubic | $x^3 - 1$ |
| 4 | quartic | $-7x^4 - x^3 + 2x^2 + 5x - 3$ |
| 5 | quintic | $x^5 - x^2 + x$ |

The classifications in Table 1-1 are not the only ones; additional names exist for polynomials of higher degree, but these are the most commonly used.

### Adding and subtracting polynomials

Remember, you can only add or subtract radicals that contain the exact same radicand and index. Similarly, you can only add or subtract polynomial terms that contain the same variables and exponents. Such terms are called **like terms**.

**Example 9:** Simplify the following expressions:

(a) $(x^3 - 6x^2 + 3x + 4) + (4x^3 + 2x^2 - 10x - 5)$

According to the commutative and associative properties of addition, you can reorder the terms and group them differently. Rewrite them as groups of like terms:

$$(x^3 + 4x^3) + (-6x^2 + 2x^2) + (3x - 10x) + (4 - 5)$$

$$5x^3 - 4x^2 - 7x - 1$$

(b) $(x^2 + 2x + 1) - 2(x + 6)$

Use the distributive property to simplify before combining like terms:

$$(x^2 + 2x + 1) - 2x - 12$$

$$x^2 + (2x - 2x) + (1 - 12)$$

$$x^2 - 11$$

## Multiplying polynomials

Terms need not be like terms in order to multiply them together. In fact, to multiply polynomials together, all you need is the distributive property.

**Example 10:** Multiply the following expressions and simplify your answer:

(a) $(2x^2 + 1)(x - 3)$

You may use the **FOIL method** to find the product. *FOIL* is a mnemonic device meaning "First, Outside, Inside, Last," describing which terms must be multiplied together. Figure 1-2 explains what is meant by each of the four groups in *FOIL*.

**Figure 1-2** Each letter in FOIL stands for a pair of terms that must be multiplied together.

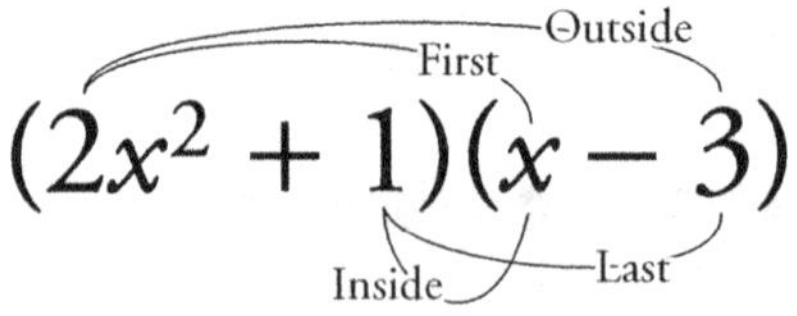


Multiply the first terms together, then the outer, the inner, and the last terms, and add the products where possible.

$$(2x^2 \cdot x) + (2x^2 \cdot -3) + (1 \cdot x) + (1 \cdot -3)$$
$$2x^3 - 6x^2 + x - 3$$

(b) $(x - 2)(x^2 - 4x + 5)$

In order for you to use the FOIL method, both polynomials must be binomials, and that's not the case here. When FOIL fails, simply distribute each term in the first polynomial into the second polynomial:

$$x(x^2 - 4x + 5) - 2(x^2 - 4x + 5)$$
$$x^3 - 4x^2 + 5x - 2x^2 + 8x - 10$$
$$x^3 - 6x^2 + 13x - 10$$

You will review how to divide polynomials in Chapter 3.

## Rational Expressions

Just as any number that can be expressed as a fraction is called a *rational number*, any expression written as a fraction is called a **rational expression**. Operations on rational expressions follow the same governing rules as operations on fractions.

### Adding and subtracting rational expressions

All fractions must have common denominators before they can be combined via addition or subtraction.

**Example 11:** Simplify the expression:

$$\frac{3}{x} + \frac{x-1}{x+2} - \frac{x}{x-5}$$

The **least common denominator** (LCD) for this expression is $x(x + 2)(x - 5)$, since that is the smallest expression containing one of each of the pieces of every denominator. Multiply each fraction by the values necessary to get that denominator. Remember, multiplying by a fraction which has the same numerator and denominator is the same as multiplying by 1, so you're not changing the values of the original fractions.

$$\frac{3}{x} \cdot \frac{(x+2)(x-5)}{(x+2)(x-5)} + \frac{x-1}{x+2} \cdot \frac{x(x-5)}{x(x-5)} - \frac{x}{x-5} \cdot \frac{x(x+2)}{x(x+2)}$$
$$\frac{3\left(x^2-3x-10\right)}{x(x+2)(x-5)} + \frac{(x-1)\left(x^2-5x\right)}{x(x+2)(x-5)} - \frac{x^2(x+2)}{x(x+2)(x-5)}$$

Now, all the numerators can be written over the LCD.

$$\frac{\left(3x^2-9x-30\right)+\left(x^3-6x^2+5x\right)-\left(x^3+2x^2\right)}{x(x+2)(x-5)}$$

$$\frac{-5x^2-4x-30}{x^3-3x^2-10x}$$

## Multiplying rational expressions

You do not need common denominators in order to multiply rational expressions. To find the product of two fractions, simply multiply the numerator of the first times the numerator of the second; then do likewise with the denominators.

**Example 12:** Simplify the expression:

$$\left(\frac{x+3}{5}\right)\left(\frac{x^2-1}{2x}\right)$$

$$\frac{(x+3)\left(x^2-1\right)}{5\cdot 2x}$$

$$\frac{x^3+3x^2-x-3}{10x}$$

## Simplifying complex fractions

When fractions are divided, the result is a **complex fraction**, a fraction that itself contains fractions. To simplify such fractions, you will employ a method that changes the division problem into multiplication.

**Example 13:** Simplify the complex fraction:

$$\frac{\frac{x+1}{x}}{\frac{x^2}{2x-3}}$$

Begin by rewriting the complex fraction as a division problem:

$$\frac{x+1}{x}\div\frac{x^2}{2x-3}$$

Take the reciprocal of the second fraction (turn it upside down) and change the division sign to multiplication:

$$\frac{x+1}{x}\cdot\frac{2x-3}{x^2}$$

Multiply as you would ordinary fractions:

$$\frac{2x^2 - x - 3}{x^3}$$

## Equations and Inequalities

Your primary task as a precalculus student will be to solve equations and inequalities using a variety of techniques, so it's worthwhile to make sure you have a good mastery of the techniques you should know thus far.

### Solving equations

To solve an equation for a variable (that is, to isolate the variable on one side of the equal sign), you may do any of the following:

- Add or subtract the same quantity on both sides of the equal sign.
- Multiply or divide both sides of the equal sign by the same non-zero quantity. However, make sure that you do not multiply or divide by a variable quantity, if at all possible. Doing so could result in additional or lost solutions, respectively.
- Cross-multiply to eliminate fractions.

$$\frac{a}{b} = \frac{c}{d} \text{ becomes } ad = bc$$

**Example 14:** Solve the following equations:

(a) $2x - 4 = 3(x - 9) + 2$

Distribute the 3, and then move the variables to the left and the constants to the right sides of the equation.

$$2x - 4 = 3x - 27 + 2$$
$$-x = -21$$

Divide both sides by $-1$ to get the answer: $x = 21$.

(b) $\frac{x+2}{x-1} + \frac{x}{x+2} = 2$

Subtract the second fraction from both sides:

$$\frac{x+2}{x-1} = 2\left(\frac{x+2}{x+2}\right) - \frac{x}{x+2}$$
$$\frac{x+2}{x-1} = \frac{x+4}{x+2}$$

Cross-multiply and solve:

$$(x+2)(x+2)=(x-1)(x+4)$$

$$x^2+4x+4=x^2+3x-4$$

$$x=-8$$

(c) $|3x+1|=7$

If only the quantity in absolute values appears on the left side of the equation, you can rewrite it as two different equations, both without absolute value signs. In one, you simply set the sides equal; in the other, the right-hand side of the equation is written as its opposite:

$$3x+1=7 \text{ or } 3x+1=-7$$

$$3x=6 \text{ or } 3x=-8$$

$$x=2 \text{ or } -\frac{8}{3}$$

## Solving linear inequalities

Linear inequalities are treated almost exactly like equations. The only difference is that multiplying or dividing both sides of the inequality by a negative value reverses the inequality sign. For example, ≥ becomes ≤ and < becomes >.

**Example 15:** Give the solutions in interval notation:

(a) $3x+4<5x+7(x-1)$

Distribute the 7 and isolate the $x$ terms as if this were an equation:

$$3x+4<5x+7x-7$$

$$3x+4<12x-7$$

$$-9x<-11$$

To finish, you have to divide by –9, so reverse the inequality symbol:

$$x>\frac{11}{9}$$

In interval form, the answer is $\left(\frac{11}{9},\infty\right)$

(b) $-5\le 2x+3<13$

Subtract 3 from all parts of the inequality, and then divide everything by 2 to isolate the $x$:

$$-8 \le 2x < 10$$

$$-4 \le x < 5$$

The answer, in interval form, is $[-4,5)$.

## Solving absolute value inequalities

Just as absolute value equations require you to solve two equations, absolute value inequalities require you to solve two inequalities. The procedures are different for problems involving less-than signs and those involving greater-than signs.

**Example 16:** Give the solutions in interval notation:

(a) $|x-4|-3<6$

Isolate the absolute value quantity on the left side:

$$|x-4|<9$$

Transform this into a double inequality, removing the absolute value signs. The quantity on the far left will be the opposite of the quantity on the far right, and the inequality signs match:

$$-9<x-4<9$$

Solve this just like Example 15(b):

$$-5<x<13$$

The answer is $(-5,13)$.

(b) $|3x+1|\ge 4$

Inequalities involving the greater-than symbol must be rewritten as two linear inequalities. In the first, simply drop the absolute value signs. In the second, reverse the sign and change the constant on the right to its opposite:

$$3x+1\ge 4 \quad \text{or} \quad 3x+1\le -4$$

$$3x\ge 3 \quad \text{or} \quad 3x\le -5$$

$$x\ge 1 \quad \text{or} \quad x\le -\frac{5}{3}$$

In interval form, the answer is $[1,\infty)$ or $(-\infty,-\frac{5}{3}]$. You can replace the word "or" with the symbol $\cup$; both notations are correct.

## Special inequality cases

Whenever you are presented with inequality problems containing rational expressions or polynomials of a degree higher than one (such as quadratics or cubics), you must use an altogether different method. Here are the steps you should follow.

1. Move all terms to the left side of the inequality, leaving only 0 on the right side.
2. Find the **critical numbers**, the values for which the left side of the inequality either equals 0 or is undefined. (Remember, a fraction equals 0 when its numerator equals 0 and is undefined when its denominator equals 0.)
3. Draw a number line and mark the critical points on it. Use a closed dot to represent included points (points that could be a solution) and an open dot to represent unattainable points (such as places where the expression is undefined or where the inequality sign does permit the possibility of equality).
4. Treat those dots as boundaries that split the number line into intervals and choose one value (called a **test point**) from each segment, *between the critical numbers.*
5. Each interval whose test point makes the original inequality true is a solution.

**Example 17:** Give the solutions in interval notation:

(a) $\frac{x+1}{x-2} \geq 3$

Subtract 3 from both sides and simplify:

$$\frac{x+1}{x-2} - 3 \cdot \frac{x-2}{x-2} \geq 0$$

$$\frac{x+1-3x+6}{x-2} \geq 0$$

$$\frac{-2x+7}{x-2} \geq 0$$

The numerator equals 0 when $x = \frac{7}{2}$, and the denominator equals 0 when $x = 2$; both are critical numbers. Since the inequality includes the possibility of equality, you use a solid dot for $\frac{7}{2}$. However, 2 makes the fraction undefined and cannot be a solution; use an open dot for it, as shown in Figure 1-3.

**Figure 1-3** The critical numbers 2 and $\frac{7}{2}$ break the number line into three distinct intervals.

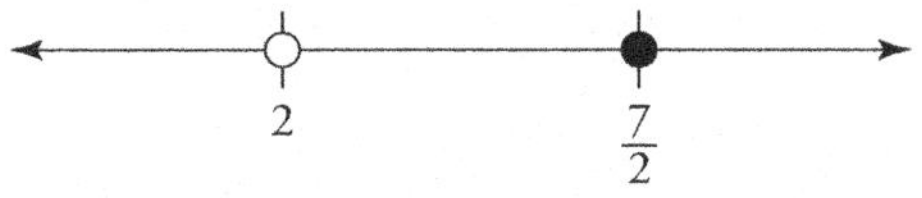


The numbers $x = 0$, 3, and 5 belong to the pictured intervals, from left to right. When you plug each into the original inequality, only $x = 3$ makes it true. Therefore, the interval in which it belongs is the solution: $(2, \frac{7}{2}]$.

(b) $2x^2 - 5x - 3 < 0$

Wherever the trinomial equals 0, place an open dot critical point on the number line. You need to factor in order to find these values. (A brief review of factoring is given in Chapter 3.)

$$(x - 3)(2x + 1)$$

Set both factors equal to 0 to get critical numbers of $x = 3$ and $-\frac{1}{2}$. Choose test values from the resulting intervals of $(-\infty, -\frac{1}{2})$, $(-\frac{1}{2}, 3)$ and $(3, \infty)$, and test them in the original inequality. The correct answer is $(-\frac{1}{2}, 3)$.

## Finding Linear Equations

You need only two items to write the equation of any line: its **slope** (a fraction describing how quickly the line rises vertically compared to how it rises horizontally) and any point on the line. Once you have that information, plug it into the correct spots of **point-slope form** for a linear equation:

$$y - y_1 = m(x - x_1)$$

where $m$ is the slope and the point you were given on the line is $(x_1, y_1)$.

If you do not know the slope of the line but are given two points, $(x_1, y_1)$ and $(x_2, y_2)$, you can calculate the slope using this equation:

$$m = \frac{y_2 - y_1}{x_2 - x_1}$$

**Example 18:** Find the equations of the following lines:

(a) line *l*, which has slope –2 and passes through point (–1,5)

Set $m = -2$, $x_1 = -1$, and $y_1 = 5$; plug these values into point-slope form:

$$y - y_1 = m(x - x_1)$$
$$y - 5 = -2(x - (-1))$$
$$y - 5 = -2x - 2$$

If you solve this equation for $y$, you get the **slope-intercept form** for a line ($y = mx + b$), where $m$ is once again the slope and $b$ is the line's $y$-intercept:

$$y = -2x + 3$$

(b) line *k*, which passes through points (–2,6) and (3,–5)

Begin by calculating the slope:

$$m = \frac{y_2 - y_1}{x_2 - x_1}$$
$$= \frac{-5 - 6}{3 - (-2)}$$
$$= -\frac{11}{5}$$

Now use point-slope form with either of the given points:

$$y - y_1 = m(x - x_1)$$
$$y - 6 = -\frac{11}{5}(x - (-2))$$
$$y - 6 = -\frac{11}{5}x - \frac{22}{5}$$
$$y = -\frac{11}{5}x + \frac{8}{5}$$

No matter which point you choose when plugging into point-slope form, you'll get the identical answer when you solve for $y$ and express your answer in slope-intercept form.

(c) line *n*, which has $y$-intercept –3 and is parallel to $y = 2x + 1$

Lines that are parallel have equal slopes, so the slope of line *n* will be 2. (Perpendicular lines have slopes that are negative reciprocals.) Because you already know the $y$-intercept for line *n*, use slope-intercept form: $y = 2x - 3$.

## Chapter Checkout

### Q&A

1. True or False: All rational numbers are also real numbers.
2. Express the inequality $x \geq 7$ in interval notation.
3. True or False: (1 + 2) + 3 = 3 + (1 + 2) because of the associative property of addition.
4. Simplify this radical: $\sqrt{72x^3y}$.
5. Express the solution in interval notation: $|x - 2| + 3 > 5$.
6. Express the solution in interval notation: $x^2 \leq 9$.
7. Find the equation of the line through points (0,–3) and (–2,7), and write the linear equation in slope-intercept form.

**Answers: 1.** T **2.** $[7,\infty)$ **3.** F **4.** $6|x|\sqrt{2xy}$ **5.** $(-\infty,0)\cup(4,\infty)$ **6.** $[-3,3]$
**7.** $y = -5x - 3$

# Chapter 2
# FUNCTIONS

## Chapter Check-In

- ❑ Differentiating between relations and functions
- ❑ Understanding domain and range
- ❑ Introducing important functions and their graphs
- ❑ Stretching, shifting, and reflecting function graphs
- ❑ Combining multiple functions
- ❑ Designing inverse functions

The majority of the equations you deal with in precalculus are functions. Functions are equations with specific properties, and these properties allow many freedoms. In this chapter, you learn what functions are, how to recognize them, how to graph them, and then how to manipulate them.

## Relations vs. Functions

A function is a special kind of relation. Therefore, before you can understand what a function is, you must first understand what relations are.

### Understanding relations

A **relation** is a diagram, equation, or list that defines a specific relationship between groups of elements. This is a relatively formal definition for a very basic concept. Consider the relation $r$ defined as:

$$r{:}\{(-1,3), (0,9), (1,5), (2,7), (3,2)\}$$

Here, $r$ expresses a relationship among five pairs of numbers; each pair is defined by a separate set of parentheses. Think of each set of parentheses as an (*input, output*) pairing; in other words, the first number in each pair represents the input, and the second number is the output $r$ gives for that input. For example, if you input the number –1 into $r$, the relation gives an output of 3, since the pair (–1,3) appears in the definition of $r$.

The relation $r$ is not designed to accept all real numbers as potential inputs. In fact, it will accept inputs only from the set {–1, 0, 1, 2, 3}; these numbers are the first piece of each pair in the definition of $r$. That set of potential inputs is called the **domain** of $r$. The **range** of $r$ is the set of possible outputs (the second number from each of the pairings): {2, 3, 5, 7, 9}. It is customary to order the sets from least to greatest.

## Defining functions

A **function** is a relation whose every input corresponds with a single output. This is best explained visually. In Figure 2-1, you see two relations, expressed as diagrams called **relation maps**. Both have the same domain, {$A$, $B$, $C$, $D$}, and range, {1, 2, 3}, but relation $g$ is a function, while $h$ is not.

**Figure 2-1** Two relations, $g$ and $h$, look very similar, but $g$ is a function and $h$ is not. To see why, examine the mapping paths that lead from $B$ in the relations.

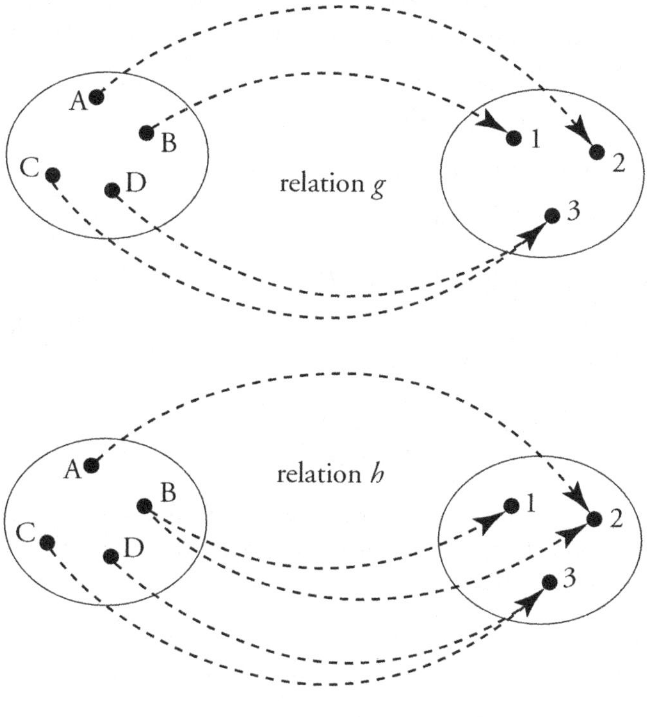

Notice that in $h$ the input $B$ is paired with two different outputs, both 1 and 2. This is not allowed if $h$ is to be a function. To be a function, each input is allowed to pair with only one output element. Visually, there can be only one path leading from each member of the domain to a member of the range. You may have noticed that in both relations shown in Figure 2-1, the inputs $C$ and $D$ result in the same output, 3. That is allowed for functions; two roads may lead *to* a single *output*, but two roads cannot lead *from* a single *input*.

A special term is reserved for a function in which every output is the result of a unique input. That is to say, there is only one road leading out from each input and only one road leading into each output. Those functions are said to be **one-to-one**.

### Writing functions

If all relations were written as ordered pair or visual maps, it would be simple to tell which of them were functions. However, it would also be tedious and inconvenient to write functions that had more than a handful of domain and range elements. Therefore, most functions are written using **function notation**. Take, for example, the function $y = x^2$. You know that $y$ is a function of $x$ because for every number $x$ you plug into $x^2$, you can get only one corresponding output. Written in function notation, that function looks like $f(x) = x^2$.

*Function notation* is handy for two reasons:

- It contains the name of the function
- It's easy to tell the value you're plugging into the function

**Example 1:** Evaluate the function $f(x) = 2x^2 + x - 3$ for $x = -1$.

Evaluating the function at $x = -1$ is the same as finding the value $f(-1)$. Plug in $-1$ everywhere you see an $x$:

$$f(-1) = 2(-1)^2 + (-1) - 3$$
$$f(-1) = 2 - 4 = -2$$

Occasionally, you'll encounter **piecewise-defined** functions. These are functions whose defining rules change based on the value of the input, and are usually written like this:

$$f(x) = \begin{cases} g(x), & x < a \\ h(x), & x \geq a \end{cases}$$

In $f(x)$, any input that is less than the value $a$ must be plugged into $g$. For instance, if $c < a$, then $f(c) = g(c)$. On the other hand, if your input is greater than or equal to $a$, $h(x)$ gives you the correct output for $f$. Remember that the inequality restrictions are based on the number you *input*, not the output of the function.

**Example 2:** Find the following values for

$$g(x) = \begin{cases} x+6, & x \le 1 \\ x-3, & x > a \end{cases}$$

(a) $g(-2)$

The defining rule for $g$ changes from $x + 6$ to $x - 3$ once your input is greater than 1. However, because the input is –2, you should stick with the first rule: $x + 6$.

$$g(-2) = x + 6 = (-2) + 6 = 4$$

(b) $g(1)$

Note that $g$ gets its value from the expression $x + 6$ when the input is less than *or equal to 1*:

$$g(1) = x + 6 = 1 + 6 = 7$$

(c) $g(5)$

Now that if the input is greater than 1, you use $x - 3$ to get the value for $g$:

$$g(5) = x - 3 = 5 - 3 = 2$$

You now know enough to determine whether given relations possess the proper characteristics to be classified as functions.

**Example 3:** Explain why, in each of the following relations, $y$ is *not* a function of $x$.

(a) $x^2 + y^2 = 9$

Begin by solving for $y$:

$$y^2 = 9 - x^2$$
$$y = \pm\sqrt{9 - x^2}$$

Notice that any valid input for $x$ (except for $x = -3$, 0, and 3) will result in *two* corresponding outputs. For example, if $x = 2$, then

$$y = \pm\sqrt{9 - (2)^2}$$
$$y = +\sqrt{5}, -\sqrt{5}$$

Remember, functions can allow only one output per input.

(b) $y = \begin{cases} x^2 + 3, & x \leq 0 \\ x^2 - 1, & x \geq 0 \end{cases}$

When $x = 0$, this function has two outputs. Notice that both conditions in the piecewise definition include 0, so $y = 3$ and $-1$ when $x = 0$. Because one input cannot have two corresponding outputs, this is not a function.

## Function Graphs

The simplest, although most labor intensive, way to graph any function is to plug in many input values to see what results and plot the (*input, output*) pairs on the coordinate plane. In fact, if you want to produce a relatively exact, hand-drawn graph, this is your only alternative once the functions become more complex than simple linear equations. In Figure 2-2, five different $x$-values provide a pretty good graph of the function $f(x) = x^2$.

**Figure 2-2** Whether a simple or complex function, the one sure way to get an accurate graph is to evaluate that function at a number of points.

| $x$ | $f(x) = x^2$ | point |
|---|---|---|
| $-2$ | $f(-2) = (-2)^2 = 4$ | $(-2, 4)$ |
| $-1$ | $f(-1) = (-1)^2 = 1$ | $(-1, 1)$ |
| 0 | $f(0) = (0)^2 = 0$ | $(-0, 0)$ |
| 1 | $f(1) = (1)^2 = 1$ | $(1, 1)$ |
| 2 | $f(2) = (2)^2 = 4$ | $(2, 4)$ |

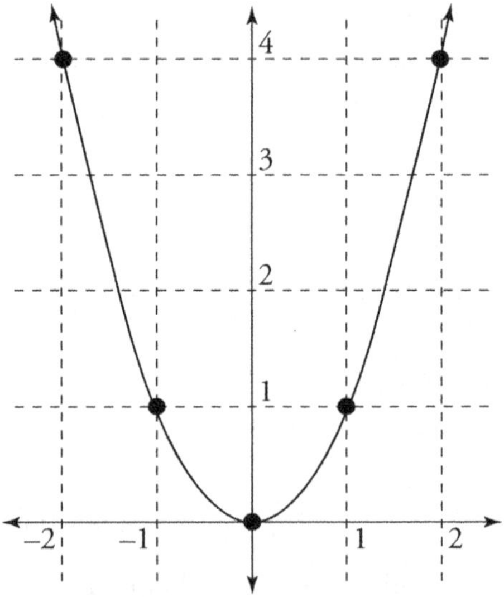


You will learn more advanced methods of creating quicker, but slightly less accurate, graphs later in this chapter. Before that, you should familiarize yourself with some important information about the graphs of functions.

## The vertical line test

You can quickly determine whether a given graph is, in fact, the graph of a function by using the **vertical line test**. If a vertical line can be drawn through the graph and that line intersects the graph in more than one location, then the graph cannot be that of a function. In Figure 2-3, you see the graph of a relation named *r*, which cannot be a function since the vertical line drawn at $x = c$ intersects the graph in two places: $(c,a)$ and $(c,b)$.

**Figure 2-3** The line $x = c$ is just one of many vertical lines that can be drawn through *r*, intersecting it in more than one place.

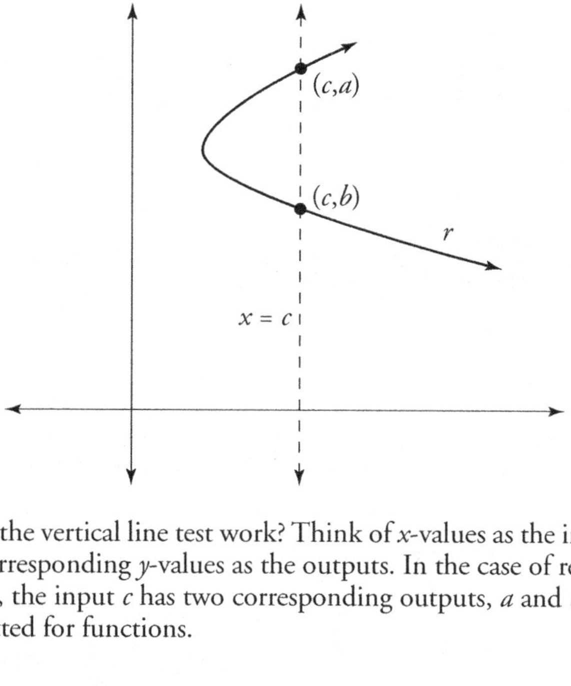


How does the vertical line test work? Think of *x*-values as the inputs for *r* and the corresponding *y*-values as the outputs. In the case of relation *r* in Figure 2-3, the input *c* has two corresponding outputs, *a* and *b*, which is not permitted for functions.

## Finding symmetry

Functions may exhibit *y-symmetry* or *origin-symmetry*. A graph exhibiting **symmetry** contains parts that will mirror one another in some fashion. Specifically, if a function $f(x)$ has a *y*-symmetric graph, then for every point $(x,y)$ on the graph, you'll also find $(-x,y)$. In the case of origin symmetry, if the graph contains the point $(x,y)$, then it must also contain $(-x,-y)$, as shown in Figure 2-4.

**Figure 2-4** A $y$-symmetric graph, like $f(x)$, looks the same on either side of the $y$-axis. An origin-symmetric graph, like $g(x)$, acts in a completely opposite manner on either side of the origin. If one side goes up and to the right, the other goes down and left.

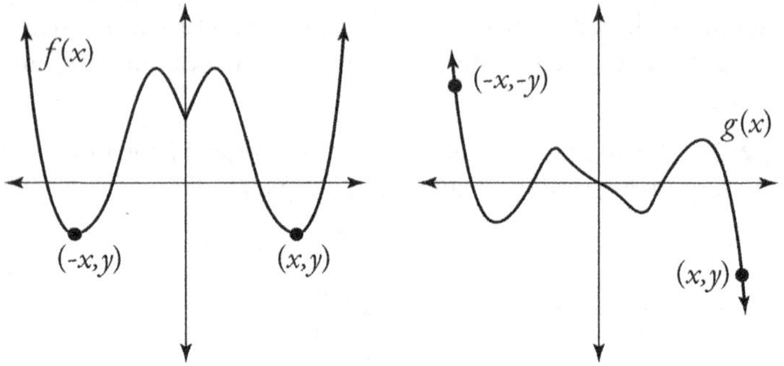


**Example 4:** Show that the graphs of the following functions demonstrate the indicated type of symmetry.

(a) $h(x) = x^4 + x^2 - 2$, $y$-symmetry

Begin by plugging in $-x$ for each $x$.

$$h(-x) = (-x)^4 + (-x)^2 - 2$$
$$h(-x) = x^4 + x^2 - 2$$

The even exponents caused the $-x$'s to become positive, and you find that $h(-x) = h(x)$. When this is true, $h$ must be $y$-symmetric.

(b) $j(x) = x^5 + x^3$, origin-symmetry

Replace all $x$'s with $-x$'s, and replace $j(x)$ with $-j(x)$:

$$-j(-x) = (-x)^5 + (-x)^3$$
$$-j(-x) = -x^5 - x^3$$

Divide everything by $-1$:

$$j(-x) = x^5 + x^3$$

Since the right side matches the original function, $j$ must be origin-symmetric.

Note that origin-symmetric functions are also called **odd functions** (since they usually contain only odd-powered exponents); $y$-symmetric functions are called **even functions** for the same reason.

## Calculating intercepts

The **intercepts** of a function are the real number values at which the graph crosses either the *x*- or the *y*-axis. Whereas a function may have numerous *x*-intercepts, it may have only one *y*-intercept to pass the vertical line test. The *x*-intercepts are also known as the function's **roots** or **zeros**.

**Example 5:** Find the intercepts of the function $f(x) = x^2 - 7x + 12$.

To calculate the *y*-intercept, substitute 0 for *x*:

$$f(0) = 0^2 - 7(0) + 12 = 12$$

The *y*-intercept is 12, since $f(x)$ crosses the *y*-axis at the point (0,12). To find the *x*-intercepts, substitute 0 for $f(x)$:

$$x^2 - 7x + 12 = 0$$
$$(x - 3)(x - 4) = 12$$
$$x = 3, 4$$

The graph of *f* will intercept the *x*-axis twice, at $x = 3$ and $x = 4$. If you can't solve the quadratic equation above, review the process in Chapter 3.

## Determining domain and range

In most cases, it's very simple to determine the domain and range of a function based solely on its graph. Because the domain represents the set of inputs, or *x*'s, if a portion of the graph appears above or below any value on the *x*-axis, that number must be in the domain of the function. Similarly, if a piece of the graph appears to the left or right of any value on the *y*-axis, that value must appear in the range of the function.

**Example 6:** Given the graph (Figure 2-5) of $h(x)$, identify the function's domain and range.

Look along the *x*-axis. Every *x*-value has a portion of the graph above or below it except for the space between $x = -1$ and 1. Since there's a dot at $x = 1$, that's included in the domain, but $x = -1$ is not. Therefore, the domain is $(-\infty, -1) \cup [1, \infty)$. Along the *y*-axis, the only values that have no portion of the graph to either side occur in the small vertical gap between $y = -2$ and $-1$. Thus, the range is $(-\infty, -2) \cup [-1, \infty)$.

**Figure 2-5** The graph of $h(x)$, used in Example 6.

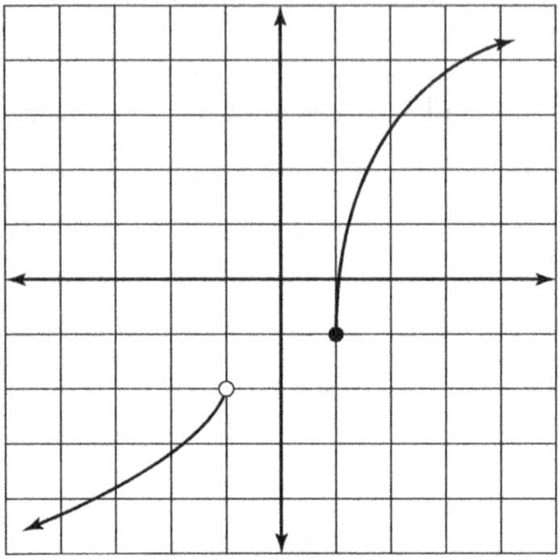

## Eight Key Function Graphs

The vast majority of the graphs you'll create are actually just transformed (stretched, shifted, and reflected) versions of the simple graphs in Figure 2-6. You should be able to recognize these graphs on sight (many of them will already be familiar to you). In addition, you should memorize the graph, key points, and domain and range of each.

- **Reciprocal function,** $y = \frac{1}{x}$. *Domain and range:* $(-\infty, 0) \cup (0, \infty)$

  Since this function outputs the reciprocal of the input, you cannot input 0, nor can you get 0 as an output. Therefore, both the lines $x = 0$ and $y = 0$ are asymptotes.

- **Quadratic graph,** $y = x^2$. *Domain:* $(-\infty,\infty)$. *Range:* $[0,\infty)$.

  The most basic parabola of all, this graph must have a nonnegative range, because the square of any number is positive.

- **Cubic graph,** $y = x^3$. *Domain and range:* $(-\infty,\infty)$.

  This graph looks a lot like $y = x^2$, but it's steeper and its left half is bent down into negative $y$ values. This makes sense, because cubics are larger than squares and can have negative outputs.

**Figure 2-6** Eight important function graphs you should memorize.

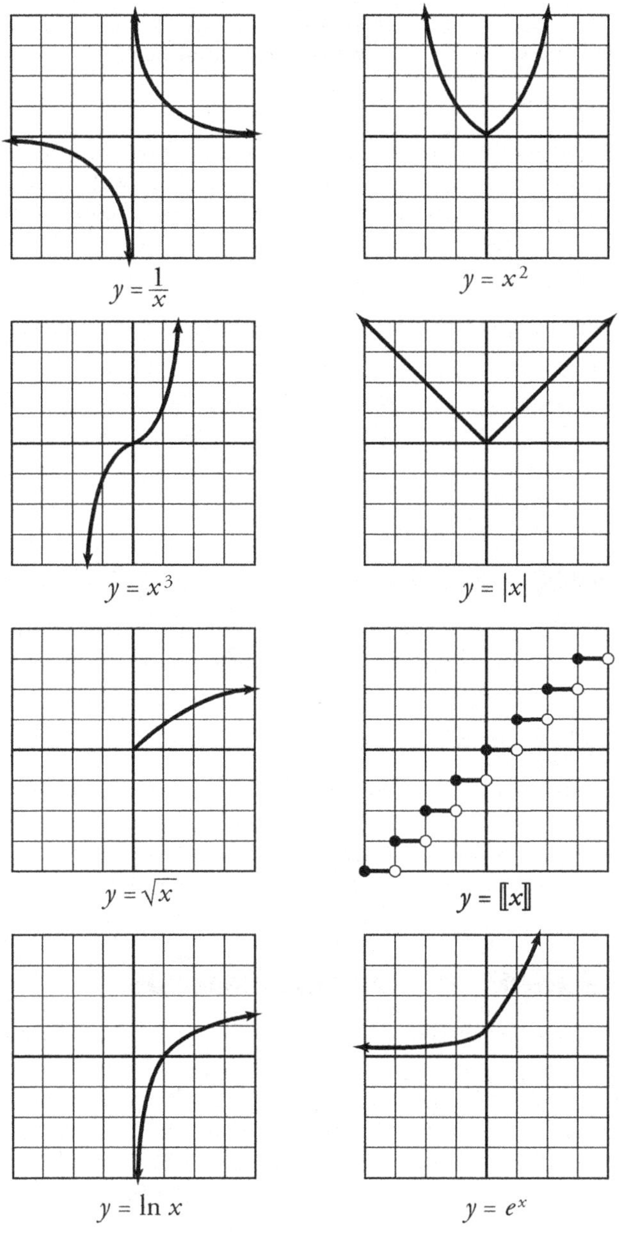

- **Absolute value,** $y = |x|$. *Domain:* $(-\infty,\infty)$. *Range:* $[0,\infty)$. The output for this function is simply the input without any negative sign. Note that it is essentially the graph of $y = x$, with its left side bent upward, so that it gives positive outputs rather than negative.

- **Square root,** $y = \sqrt{x}$. *Domain and range:* $[0,\infty)$.

  Since you can take the square root of a positive number only, and a square root always outputs a positive number, both the domain and range for this function must be non-negative. Note that the graph passes through (1,1) since the square root of 1 is 1. Similarly, it passes through (4,2), since the square root of 4 is 2.

- **Greatest integer function,** $y = [\![x]\!]$. *Domain:* $(-\infty,\infty)$. *Range: all integers.*

  This function returns the largest integer *less than or equal to* the input. Therefore, $[\![5.95]\!] = 5$, since 5 is the largest integer which is less than or equal 5.95. Negatives are a little tricky; notice that $[\![-1.6]\!] = -2$, not $-1$, so you don't just drop the decimal portion when you calculate the greatest integer output. The output is $-2$ (the largest integer *less than* $-1.6$). (Remember, $-1 > -1.6$.)

- **Natural logarithmic function,** $y = \ln x$. *Domain:* $(0,\infty)$. *Range:* $(-\infty,\infty)$.

  You'll learn more about this function in Chapter 4. For now, remember that it accepts only positive inputs, contains the point (1,0), and increases very slowly as $x$ gets larger.

- **Natural exponential function,** $y = e^x$. *Domain:* $(-\infty,\infty)$. *Range:* $(0,\infty)$.

  Again, Chapter 4 provides you with a better understanding of this function. Until then, know that the function outputs only positive numbers (although it accepts any real number input), it contains the point (0,1), and the function increases very quickly as $x$ gets larger.

## Basic Function Transformations

Just by adding a well-placed constant, negative sign, or set of absolute value symbols, you can contort, twist, and bend graphs in almost any way imaginable. The practical application of this knowledge is a technique that makes creating quick sketches of graphs an almost trivial matter. Whereas the task of graphing the function $f(x) = -2(x + 3)^2 + 1$ might at first seem complicated, it becomes much easier once you see $f$ as a series of transformations on the much simpler graph of $y = x^2$.

### Vertical and horizontal shifts

Adding to or subtracting a real number *from* a function or *within* that function causes its entire graph to shift either vertically or horizontally. Specifically, the graph of $y = f(x) + a$ is just the graph of $y = f(x)$ moved up $a$ units (if $a > 0$), or moved down $a$ units (if $a < 0$). If, however, you add or subtract that value *within* the function, the graph shifts horizontally. So, the graph of $y = f(x + b)$ is the just the entire graph of $y = f(x)$ moved $b$ units to the right (if $b < 0$), or $b$ units to the left (if $b > 0$).

Note that horizontal shifts work differently than vertical shifts. In the previous description, a positive $a$ value moves the graph up, so you might assume that a positive $b$ value moves the graph to the right. It does not; instead, the graph moves left.

**Example 7:** Sketch the graph of $f(x) = (x + 2)^2 - 3$.

You already know that the graph of $y = x^2$ is a parabola whose vertex lies on the origin. In the function $f$, you are adding 2 *within the squared function* and subtracting 3 *from the squared function*. Therefore, the 2 corresponds to a leftward horizontal shift (since 2 is positive) and the 3 equates to a downward vertical shift. See Figure 2-7 for the graph.

**Figure 2-7** The graph of $f(x)$ is simply the graph of $y = x^2$ moved 2 units left and 3 units down.

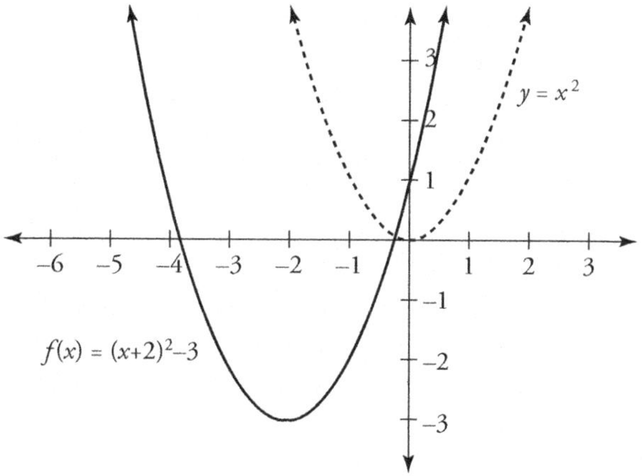

## Reflections

There are two important kinds of graph reflections, and both are the result of multiplying by –1:

- The graph of $-f(x)$ is the graph of $f(x)$ reflected across the $x$-axis. In other words, all the original points $(x,y)$ become $(x,-y)$. For an example, see Figure 2-8.

**Figure 2-8** The graphs of $f_1(x) = \ln x$ and $f_2(x) = -\ln x$ are reflections of each other across the $x$-axis.

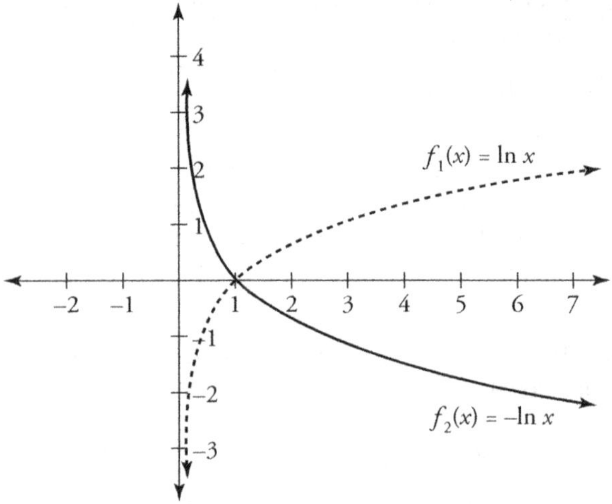


- The graph of $f(-x)$ is the graph of $f(x)$ reflected across the $y$-axis. In this case, all the original points $(x,y)$ become $(-x,y)$. For an example, see Figure 2-9.

**Figure 2-9** The graphs of $g_1(x) = \sqrt{x}$ and $g_2(x) = \sqrt{-x}$ are reflections of each other across the *y*-axis.

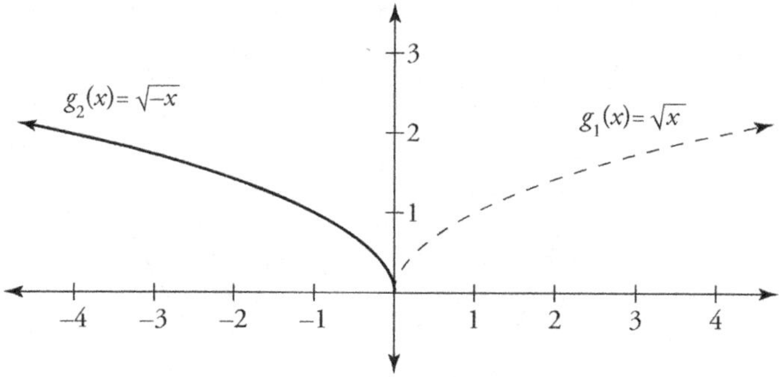


## Stretching and squishing

If you multiply an entire function by a positive constant *a*, you are essentially multiplying all of the outputs, or *y*-values, by *a*. If $a > 1$, this will stretch the graph, making the positive heights more positive and the negative heights more negative. If, on the other hand, $0 < a < 1$, the heights of each point on the graph will squish, getting closer to the *x*-axis.

In Figure 2-10, you'll find three graphs: $h(x) = x^2$ and two of its transformations. Note how the graph of $2h(x)$ is stretched, compared to the original graph; all of the functions' heights are twice as far away from the *x*-axis than they were originally. On the other hand, $\frac{1}{2}h(x)$ possesses function heights 50 percent smaller than $h(x)$.

**Figure 2-10** The graph of $a \cdot h(x)$ will stretch or squish the graph of *a* depending on *a*'s value.

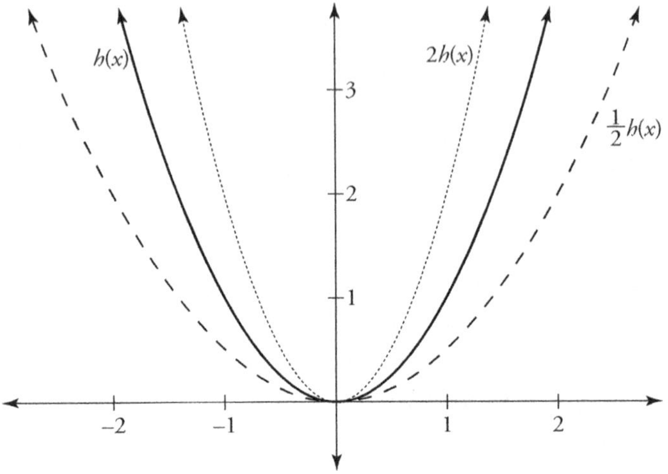


## Multiple transformations

If more than one transformation is applied to a function, this is the order you should follow when sketching the new graph:

1. Reflections
2. Stretching or squishing
3. Vertical and horizontal shifts

**Example 8:** Sketch the graph of $f(x) = -3|x - 1| + 2$.

Take the basic graph of $y = |x|$, flip it across the *x*-axis, stretch it to three times its normal height, and then move it 1 unit right and 2 units up, as shown in Figure 2-11.

**Figure 2-11** The graph of $f(x)$, the answer to Example 8.

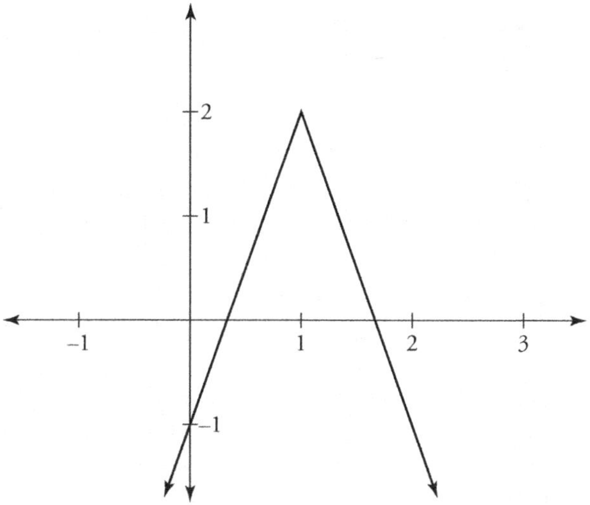


## Combining and Composing Functions

You can create new functions by combining existing functions. Usually, these new functions are the result of something as simple as addition or subtraction, but functions are capable of combining in ways other than those simple binary operations.

### Arithmetic combinations

First let's look at the easiest way to create a new function from existing functions: performing basic arithmetic operations.

**Example 9:** If $f(x) = x^2 - 2x - 3$ and $g(x) = x + 1$, find the following.

(a) $(f - g)(-1)$

Subtract $g(x)$ from $f(x)$ and substitute $-1$ for $x$:

$$(f - g)(x) = f(x) - g(x) = (x^2 - 2x - 3) - (x + 1) = x^2 - 3x - 4$$

$$(f - g)(-1) = (-1)^2 - 3(-1) - 4 = 1 + 3 - 4 = 0$$

(b) $\left(\frac{f}{g}\right)(x)$

Divide $f(x)$ by $g(x)$. Note that $f$ can be factored, so the fraction can then be simplified.

$$\left(\frac{f}{g}\right)(x) = \frac{x^2 - 2x + 3}{x + 1} = \frac{(x-3)(x+1)}{(x+1)} = x - 3, \text{ if } x \neq -1$$

Note that the domain of the new function is all real numbers except for $x = -1$, which is different than the domain of both $f(x)$ and $g(x)$.

## The composition of functions

The process of plugging one function into another is called the **composition of functions**. When one function is *composed* with another, it is usually written explicitly: $f(g(x))$, which is read "$f$ of $g$ of $x$." In other words, $x$ is plugged into $g$, and that result is in turn plugged into $f$. Function *composition* can also be written using this notation: $(h \circ k)(x)$, which is the mathematical equivalent of the statement $h(k(x))$.

**Example 10:** If $f(x) = x^2 + 10$ and $g(x) = \sqrt{x-1}$, find the following.

(a) $f(g(x))$

Substitute the radical representing $g(x)$ for the $x$ in $f(x)$:

$$f(g(x)) = f\left(\sqrt{x-1}\right) = \left(\sqrt{x-1}\right)^2 + 10$$
$$f(g(x)) = x - 1 + 10 = x + 9$$

(b) $(g \circ f)(4)$

This means the same thing as $g(f(4))$. First find $g(f(x))$:

$$g(f(x)) = g(x^2 + 10) = \sqrt{(x^2 + 10) - 1} = \sqrt{x^2 + 9}$$

Now you can evaluate $g(f(4))$:

$$g(f(4)) = \sqrt{(4)^2 + 9} = \sqrt{16 + 9} = 5$$

(c) $(f \circ f)(x)$

Substitute $x^2 + 10$ for the $x$ in $f(x)$:

$$f(f(x)) = f(x^2 + 10) = (x^2 + 10)^2 + 10$$
$$f(f(x)) = x^4 + 20x^2 + 110$$

## Inverse Functions

You have used **inverse functions** since your first days of algebra to cancel things out. Now that you possess a higher degree of mathematical proficiency, you can better explore why and how they work.

### What is an inverse function?

The *inverse function* for $f(x)$, labeled $f^{-1}(x)$ (which is read "*f* inverse of *x*"), contains the same domain and range elements as the original function, $f(x)$. However, the sets are switched. In other words, the domain of $f(x)$ is the range of $f^{-1}(x)$, and vice versa. In fact, for every ordered pair $(a,b)$ belonging to $f(x)$, there is a corresponding ordered pair $(b,a)$ that belongs to $f^{-1}(x)$. For example, consider this function, *g*:

$$g:\{(-2,0), (1, 3), (5, 9)\}$$

The inverse function is the set of all ordered pairs reversed:

$$g^{-1}:\{(0,-2), (3,1), (9,5)\}$$

Only one-to-one functions possess inverse functions. Because these functions have range elements that correspond to only one domain element each, there's no danger that their inverses will not be functions. The **horizontal line test** is a quick way to determine whether a graph is that of a one-to-one function. It works just like the vertical line test: If an arbitrary horizontal line can be drawn across the graph of $f(x)$ and it intersects $f$ in more than one place, then $f$ cannot be a one-to-one function.

Inverse functions have the unique property that, when composed with their original functions, both functions cancel out. Mathematically, this means that $f\left(f^{-1}(x)\right)=f^{-1}\left(f(x)\right)=x$.

## Graphs of inverse functions

Since functions and inverse functions contain the same numbers in their ordered pair, just in reverse order, their graphs will be reflections of one another across the line $y = x$, as shown in Figure 2-12.

**Figure 2-12** Inverse functions are symmetric about the line $y = x$.

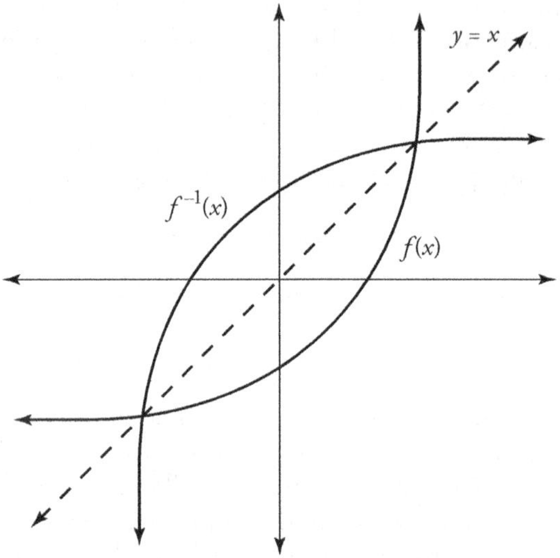


## Finding inverse functions

To find the inverse function for a one-to-one function, follow these steps:

1. Rewrite the function using $y$ instead of $f(x)$.
2. Switch the $x$ and $y$ variables; leave everything else alone.
3. Solve the new equation for $y$.
4. Replace the $y$ with $f^{-1}(x)$.
5. Make sure that your resulting inverse function is one-to-one. If it isn't, restrict the domain to pass the horizontal line test.

**Example 11:** If $f(x) = \sqrt{2x + 3}$, find $f^{-1}(x)$.

Follow the five steps previously listed, beginning with rewriting $f(x)$ as $y$:

$$y = \sqrt{2x + 3}$$

$$x = \sqrt{2y+3}$$

$$x^2 = 2y + 3$$

$$f^{-1}(x) = \frac{1}{2}(x^2 - 3),\ x \geq 0$$

Note the restriction $x \geq 0$ for $f^{-1}(x)$. Without this restriction, $f^{-1}(x)$ would not pass the horizontal line test. It obviously must be one-to-one, since it must possess an inverse of $f(x)$. You should use that portion of the graph because it is the reflection of $f(x)$ across the line $y = x$, unlike the portion on $x < 0$.

## Chapter Checkout

### Q&A

**1.** True or False: The function $k$ is one-to-one.

$$k\!:\{(-2,3),\ (-1,-2),\ (0,6),\ (1,3),\ (2,-7)\}$$

**2.** Given the function $h(x) = -(x-1)^2 + 5$:

(a) What is the domain of $h(x)$?

(b) What is the range of $h(x)$?

**3.** Identify the function $m$ shown in this graph.

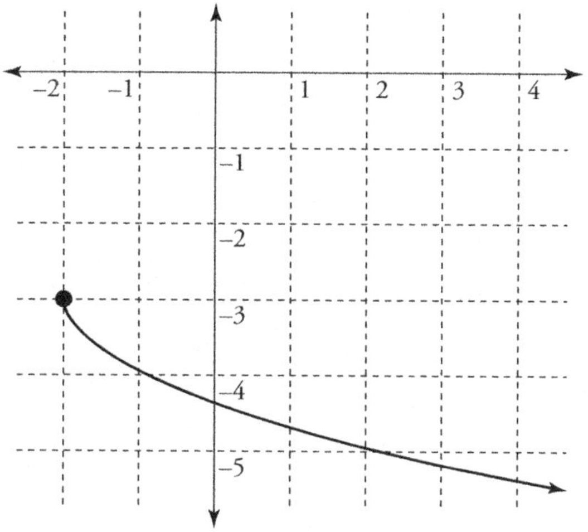

4. If $f(x) = |x - 6|$ and $g(x) = [\![x]\!]$, evaluate $(f \circ g)(-2.3)$.
5. If $j(x) = \frac{x-5}{2}$, find $j^{-1}(x)$.

**Answers: 1.** F **2. (a)** $(-\infty,\infty)$ **(b)** $(-\infty,5]$ **3.** $m(x) = -\sqrt{x+2} - 3$ **4.** 9 **5.** $j^{-1}(x) = 2x + 5$

# Chapter 3

# POLYNOMIAL AND RATIONAL FUNCTIONS

## Chapter Check-In

- ❑ Factoring polynomials
- ❑ Solving quadratic equations
- ❑ Calculating the roots of polynomials
- ❑ Dividing polynomials via long and synthetic division.
- ❑ Working with irrational roots
- ❑ Graphing rational functions

Once you can find the roots of polynomials, you possess the formidable skill to solve nearly any equation by hand. Linear equations are easy to solve, but once the polynomials increase in degree, the procedures become more involved. This chapter begins with a review of factoring and quadratic functions, and eventually graduates to more advanced topics in root calculation.

## Factoring Polynomials

**Factoring**, the process of "unmultiplying" polynomials in order to return to a unique string of polynomials of lesser degree whose product is the original polynomial, is the simplest way to solve equations of higher degree. Although you should already be proficient in factoring, here are the methods you should be familiar with, in case you need to review.

## Greatest common factors

If all of the terms in a polynomial contain one or more identical factors, combine those similar factors into one monomial, called the **greatest common factor**, and rewrite the polynomial in factored form.

**Example 1:** Factor the expressions.

(a) $15x^3 + 5x^2 - 25x$

Since each term in the polynomial is divisible by both $x$ and 5, the greatest common factor is $5x$. In factored form, the polynomial is written $5x(3x^2 + x - 5)$.

(b) $18x^3y^5z^4 + 6x^2yz^3 - 9x^2y^3z^2$

The largest monomial by which each of the terms is evenly divisible, thus the greatest common factor, is $3x^2yz^2$, so factor it out.

$$3x^2yz^2(6xy^4z^2 + 2z - 3y^2)$$

## Factoring by grouping

Sometimes, the greatest common factor of an expression is not just a monomial but an entire parenthetical quantity. You are allowed to factor out quantities in parentheses just as you can factor out individual terms.

**Example 2:** Factor the following expressions.

(a) $3x(x - 5) + 2y(x - 5) - 10(x - 5)$

The only thing that divides into each of these terms evenly is the linear expression $(x - 5)$. Factor it out, just as you would any greatest common factor, leaving behind the monomial in each term that was multiplied by $(x - 5)$:

$$(x - 5)(3x + 2y - 10)$$

(b) $3x^2 - 6x - 4x + 8$

Nothing, except the number 1, divides evenly into each of the terms, and there's no use factoring out 1. However, the first two terms have a greatest common factor of $3x$. Furthermore, if you factor $-4$ out of the final two terms, you can factor by grouping:

$$3x(x - 2) - 4(x - 2)$$

$$(x - 2)(3x - 4)$$

## Factoring quadratic trinomials

Your most common factoring task, aside from greatest common factoring, is changing a quadratic trinomial into the product of two linear binomials.

**Example 3:** Factor the following expressions.

(a) $x^2 - 4x - 12$

If the leading coefficient is 1, as it is here, the process is simple. Find two numbers whose sum equals the coefficient of the *x* term and whose product is equal to the constant term. The only two numbers whose sum is –4 and that multiply to give –12 are –6 and 2. Use these as the constants in the linear factors:

$$(x - 6)(x + 2)$$

(b) $x^2 - 10x + 24$

Since this quadratic trinomial has a leading coefficient of 1, find two numbers with a product of 24 and a sum of –10. Through some experimenting, you'll find those numbers are –6 and –4:

$$(x - 6)(x - 4)$$

(c) $2x^2 + 9x - 5$

If the leading coefficient is not 1, you must follow another procedure. You still seek two numbers, and those numbers will still add up to 9. However, they will multiply to give you –10, the product of the leading coefficient and the constant. (You didn't have to use this technique when the leading coefficient was 1, since the product of the leading coefficient and the constant would just have been the constant anyway.) The numbers in question are 10 and –1. Rewrite the *x* coefficient as the sum of those numbers:

$$2x^2 + (10 - 1)x - 5$$

Distribute the *x* to both 10 and –1:

$$2x^2 + 10x - x - 5$$

To finish, factor by grouping:

$$2x(x + 5) - 1(x + 5)$$
$$(x + 5)(2x - 1)$$

### Special factor patterns

Occasionally, the only effort you'll have to expend on a factoring problem is recognizing that the polynomial in question matches one of three specific patterns. You should memorize each of these formulas so that you can spot them instantly:

- Difference of perfect squares: $x^2 - a^2 = (x + a)(x - a)$
- Difference of perfect cubes: $a^3 - b^3 = (a - b)(a^2 + ab + b^2)$
- Sum of perfect cubes: $a^3 + b^3 = (a + b)(a^2 - ab + b^2)$

**Example 4:** Factor the following expressions completely.

(a) $27x^3 + 8$

Note that $27x^3$ and 8 are both perfect cubes, so apply the sum of perfect cubes formula:

$$27x^3 + 8 = (3x)^3 + (2)^3 = (3x + 2)[(3x)^2 - (3x)(2) + (2)^2]$$
$$(3x + 2)(9x^2 - 6x + 4)$$

(b) $20x^2 - 405$

This doesn't appear to match any of the patterns, but you can factor out a greatest common factor of 5 to begin:

$$5(4x^2 - 81)$$

Now it's clearer that $(4x^2 - 81)$ is the difference of perfect squares, since $(2x)^2 = 4x^2$ and $9^2 = 81$:

$$5(2x + 9)(2x - 9)$$

## Solving Quadratic Equations

There are three major techniques for solving quadratic equations (equations formed by polynomials of degree 2). The easiest, factoring, will work only if all solutions are rational. The other two methods, the quadratic formula and completing the square, will both work flawlessly every time, for every quadratic equation. Of those two, the quadratic formula is the easier, but you should still learn how to complete the square, because you'll need the skill again in Chapter 9.

### Factoring

To solve a quadratic equation by factoring, follow these steps:

1. Move all non-zero terms to the left side of the equation, effectively setting the polynomial equal to 0.
2. Factor the quadratic completely.
3. Set each factor equal to 0 and solve the smaller equations.
4. Plug each answer into the original equation to ensure that it makes the equation true.

**Example 5:** Solve the equation.

$$3x^3 = -13x^2 + 10x$$

Add $13x^2$ and $-10x$ to both sides of the equation:

$$3x^3 + 13x^2 - 10x = 0$$

Factor the polynomial, set each factor equal to 0, and solve.

$$x(3x - 2)(x + 5)$$

$$x = 0 \quad 3x - 2 = 0 \quad x + 5 = 0$$

$$x = 0, \frac{2}{3}, -5$$

Because all three of these $x$-values make the quadratic equation true, they are all solutions.

### The quadratic formula

If an equation can be written in the form $ax^2 + bx + c = 0$, then the solutions to that equation can be found using the **quadratic formula**:

$$x = \frac{-b \pm \sqrt{b^2 - 4ac}}{2a}$$

This method is especially useful if the quadratic equation is not factorable. A word of warning: Make sure that the quadratic equation you are trying to solve is set equal to 0 before plugging the quadratic equation's coefficients $a$, $b$, and $c$ into the formula. You should memorize the quadratic formula if you haven't done so already.

**Example 6:** Solve the quadratic equation.

$$6x = 4x^2 + 1$$

Set the equation equal to 0:

$$-4x^2 + 6x - 1 = 0$$

The coefficients for the quadratic formula are $a = -4$, $b = 6$, and $c = -1$:

$$x = \frac{-(6) \pm \sqrt{6^2 - 4(-4)(-1)}}{2(-4)}$$

$$x = \frac{-6 \pm \sqrt{20}}{-8}$$

$$x = \frac{-6 \pm 2\sqrt{5}}{-8}$$

$$x = \frac{2\left(-3 \pm \sqrt{5}\right)}{-8}$$

$$x = \frac{-3 + \sqrt{5}}{-4}, \frac{-3 - \sqrt{5}}{-4}$$

You can also write the answers as $x = \frac{3 - \sqrt{5}}{4}, \frac{3 + \sqrt{5}}{4}$, the result of multiplying the numerators and denominators of both by –1. Note that the quadratic formula technique can easily find irrational and imaginary roots, unlike the factoring method.

## Completing the square

The most complicated, though itself not very difficult, technique for solving quadratic equations works by forcibly creating a trinomial that's a perfect square (hence the name). Here are the steps to follow:

1. Put the equation in form $ax^2 + bx = c$. In other words, move only the constant term to the right side of the equation.
2. If $a \neq 1$, divide the entire equation by $a$.
3. Add the constant value $\left(\frac{b}{2a}\right)^2$ to both sides of the equation.
4. Write the left side of the equation as a perfect square.
5. Take the square roots of both sides of the equation, remembering to add the "±" symbol on the right side.
6. Solve for $x$.

**Example 7:** Solve the quadratic equation by completing the square.

$$2x^2 + 12x - 3 = 0$$

Move the constant so it alone is on the right side:

$$2x^2 + 12x = 3$$

Divide everything by the leading coefficient, since it's not 1:

$$x^2 + 6x = \frac{3}{2}$$

Half of the $x$-term's coefficient squared, $\left(\frac{6}{2}\right)^2$, = 9. Add that value to both sides of the equation:

$$x^2 + 6x + 9 = \frac{3}{2} + \frac{18}{2}$$

The left side is a perfect square:

$$(x + 3)^2 = \frac{21}{2}$$

Solve for $x$: Don't forget that you must include a ± sign when square rooting both sides of any equation.

$$\sqrt{(x+3)^2} = \pm\sqrt{\frac{21}{2}}$$

$$x = -3 \pm \sqrt{\frac{21}{2}}$$

The answer can also be written as $x = -3 \pm \frac{\sqrt{42}}{2}$, if rationalized.

## Polynomial Division

You can use one of two techniques to divide polynomials. When you found the greatest common factor, the easiest method was to see what values divided easily into every term. Likewise, you can divide polynomials to find factors of those polynomials.

### Long division

The method of *long division*, similar to the procedure used by elementary school students to divide whole numbers, is the most general method of finding the quotient of polynomials.

**Example 8:** Divide $x^4 + 3x^3 - 2x^2 + 7x + 1$ by $x^2 - 2x + 2$.

Write as a long division problem. Because you are dividing by $x^2 - 2x + 2$, it goes outside the division symbol and is called the **divisor**. The fourth-degree polynomial goes inside the division symbol, and is called the **dividend.**

$$x^2 - 2x + 2\overline{\smash{)}x^4 + 3x^3 - 2x^2 + 7x + 1}$$

Answer this question: What can you multiply the first term of the *divisor* by to get the first term of the *dividend*? In other words, $x^2$ times what equals $x^4$? The answer is $x^2$, so write that above the division symbol (in the *quotient* space), and line it up above the term $-2x^2$, since it has the same degree:

$$\begin{array}{r} x^2 \phantom{+ 7x + 1} \\ x^2 - 2x + 2\overline{\smash{)}x^4 + 3x^3 - 2x^2 + 7x + 1} \end{array}$$

Multiply the $x^2$ in the quotient by each term in the divisor and write the results below the terms in the dividend so that the degrees match. You want to subtract these terms, so change each sign to its opposite as you write it, and combine the like terms:

$$\begin{array}{r} x^2 \phantom{+ 7x + 1} \\ x^2 - 2x + 2\overline{\smash{)}x^4 + 3x^3 - 2x^2 + 7x + 1} \\ \underline{-x^4 + 2x^3 - 2x^2} \phantom{+ 7x + 1} \\ 5x^3 - 4x^2 \phantom{+ 7x + 1} \end{array}$$

Bring the next unused term in the dividend ($7x$) down, so that you now have the polynomial $5x^3 - 4x^2 + 7x$. Repeat the process, this time answering the question: What times $x^2$ equals $5x^3$ (the first term of the divisor and the first term of the new polynomial)? When finished, bring down the "+1" in the dividend and repeat the process:

$$\begin{array}{r} x^2 + 5x + 6 \\ x^2 - 2x + 2\overline{\smash{)}x^4 + 3x^3 - 2x^2 + 7x + 1} \\ \underline{-x^4 + 3x^3 - 2x^2} \phantom{+ 7x + 1} \\ 5x^3 - 4x^2 + 7x \phantom{+ 1} \\ \underline{-5x^3 + 10x^2 - 10x} \phantom{+ 1} \\ 6x^2 - 3x + 1 \\ \underline{-6x^2 + 12x - 12} \\ 9x - 11 \end{array}$$

Since $9x - 11$ has a smaller degree than the divisor, you have finished; the remainder will be $9x - 11$. The answer will be the quotient plus the remainder divided by the divisor:

$$x^2 + 5x + 6 + \frac{9x - 11}{x^2 - 2x + 2}$$

Before undertaking any long division problem, make sure there are no "missing terms" in the divisor or dividend. In other words, if the polynomial is of degree $n$, be sure it has $n + 1$ terms. For example, if you want to divide $x^3 + 2x + 6$ by $x^2 - 5$, you'll notice that the divisor has degree two, but not $2 + 1 = 3$ terms, so you need to rewrite $x^2 - 5$ as $x^2 + 0x - 5$. You'll have to do the same with the dividend, so your setup for the problem will look like this:

$$x^2 + 0x - 5 \overline{\smash{\big)}\, x^3 + 0x^2 + 2x + 6}$$

## Synthetic division

If the divisor of a polynomial division problem is linear and its leading coefficient is 1, you can use **synthetic division** to find the quotient, a method shorter than long division, using only the coefficients of the divisor and dividend.

**Example 9:** Divide $2x^3 - x^2 + 3$ by $x + 2$.

Since the divisor is in form $(x + a)$, this is a perfect candidate for synthetic division (although long division will still work). Begin by writing the opposite of the divisor's constant term (–2) in a small box, and list the coefficients of the dividend next to it. Fill in any missing terms with a 0 coefficient. (This dividend has no $x$ term, so fill in a 0 for it.) Then draw a horizontal line beneath everything, leaving some space:

$$\begin{array}{r|rrrr} -2 & 2 & -1 & 0 & 3 \\ & & & & \\ \hline & & & & \end{array}$$

Bring down the first coefficient (2) and write it below the line:

$$\begin{array}{r|rrrr} -2 & 2 & -1 & 0 & 3 \\ & \downarrow & & & \\ \hline & 2 & & & \end{array}$$

Multiply the number in the box (–2) by the number below the line (2) and write the result (–4) below the number in the next column (–1):

$$\begin{array}{r|rrrr} -2 & 2 & -1 & 0 & 3 \\ & & -4 & & \\ \hline & 2 & & & \end{array}$$

Combine the numbers in the new column (–1 – 4 = –5) and write the result below the line. Now, repeat the procedure. Multiply the number in the box (–2) by the number below the line (–5) and write the result (10) below the number in the next column (0). Continue until you have as many numbers below the line as you do coefficients in the dividend:

$$\begin{array}{r|rrrr} -2 & 2 & -1 & 0 & 3 \\ & & -4 & 10 & -20 \\ \hline & 2 & -5 & 10 & -17 \end{array}$$

The numbers below the line are the coefficients of your answer, the quotient, and its remainder. The quotient's degree will be one less than the dividend; since the dividend was a cubic, the quotient will be a quadratic. Write the remainder (–17) as a fraction divided by the divisor, just as you did with long division. The final answer is:

$$2x^2 - 5x + 10 + \frac{-17}{x+2}$$

## Important Root-Finding Theorems

Although the skill of dividing polynomials is, in and of itself, a worthwhile outcome, that skill can be applied to a greater purpose: factoring and finding the roots of polynomial functions. In this section, you'll explore the two theorems that allow you to extend the technique of synthetic division to the realm of factoring.

### The remainder theorem

The **remainder theorem** says that if a polynomial function $f(x)$ is divided by a linear term of the form $(x - a)$ and the remainder is $r$, then $f(a) = r$.

This is a startling discovery, if not altogether useful. Most of the time, evaluating a function for a constant value is not an arduous or challenging task, so you might wonder why a theorem that provides a shortcut for evaluating functions is so important. Frankly, it's not as important as its corollary, the *factor theorem*, which you'll read about in just a moment.

**Example 10:** Demonstrate the remainder theorem by showing that $f(-1)$ is equal to the remainder when the polynomial $f(x) = -x^3 + 3x^2 + x - 7$ is divided by $x + 1$.

Since the divisor is in the form $(x - a)$, use synthetic division to calculate the remainder:

| | | | | |
|---|---|---|---|---|
| $\underline{-1}\rfloor$ | $-1$ | $3$ | $1$ | $-7$ |
| | | $1$ | $-4$ | $3$ |
| | $-1$ | $4$ | $-3$ | $-4$ |

The remainder equals –4; therefore, according to the theorem, so should $f(-1)$:

$$f(-1) = -(-1)^3 + 3(-1)^2 + (-1) - 7$$
$$f(-1) = 1 + 3 - 8 = -4$$

### The factor theorem

If a non-zero number $a$ is a factor of another number $b$, when you divide $\frac{b}{a}$, you should get an integer. That is to say, $a$ divides evenly into $b$ with no remainder. For example, you know that 5 is a factor of 20 since $\frac{20}{5} = 4$, an integer with no remainder. The same is true of polynomials, as stated explicitly by the **factor theorem**. According to that theorem, a polynomial function $f(x)$ has a factor $(x - a)$ if and only if $f(a) = 0$.

In other words, if you use synthetic division to divide $f(x)$ by $(x - a)$ and get a reminder of 0, then you know that $f(a) = 0$ (according to the remainder theorem). The factor theorem takes it one step further and concludes that a remainder of 0 implies that $(x - a)$ is actually a factor of $f(x)$, meaning that you can now factor polynomials other than just quadratics or cubics that are the sum or difference of perfect squares. Remember that the middle sign of the factor will always be the opposite sign of the root itself.

**Example 11:** Prove that 4 is a root of $g(x) = x^3 - 7x^2 + 14x - 8$, and use that information to factor $g(x)$ completely.

If 4 is a root (in other words, 4 is a zero or $x$-intercept of $g$), then $g(4)$ *must* equal 0. You can show this using synthetic division:

| | | | | |
|---|---|---|---|---|
| $\underline{4}\rfloor$ | $1$ | $-7$ | $14$ | $-8$ |
| | | $4$ | $-12$ | $8$ |
| | $1$ | $-3$ | $2$ | $0$ |

Because the remainder is 0, then $g(4) = 0$, proving that 4 is a root of $g(x)$. In fact, the result of the synthetic division tells you what $g(x)$ equals when the factor $(x - 4)$ divides out evenly: $x^2 - 3x + 2$. Therefore, you know that the polynomial function $g(x)$ can be factored as

$$(x - 4)(x^2 - 3x + 2)$$

(Don't forget that the middle sign of the factor, the "–" in $x - 4$, must always be the opposite sign of the root, which in this case was +4.)

You haven't finished yet. That quadratic portion of the polynomial can be factored as $(x - 2)(x - 1)$. Therefore, the complete factorization of $g(x)$ is

$$(x - 4)(x - 2)(x - 1)$$

Now that $g(x)$ is fully factored, you can find its other two roots with no work at all. Since $(x - 2)$ and $(x - 1)$ are factors, then $x = 2$ and $x = 1$ are roots of $g$, according to the factor theorem.

## Calculating Roots

Thanks to the factor theorem, you know that the process of factoring is essentially akin to the process of finding the roots of a function, which is useful not only in equation solving, but graphing as well, since the roots of a function are also its $x$-intercepts. In this section, you explore the theory and practice of root calculation.

### The Fundamental Theorem of Algebra

A number of algebraic theorems lay the foundation and justify all of the root-finding techniques you've used so far, as well as those you use in the remainder of this chapter. The most important of these theorems is the **Fundamental Theorem of Algebra**, which guarantees that any polynomial of degree $n$ will have exactly $n$ total roots.

This doesn't mean that those roots will all be real numbers. Some may be imaginary roots, meaning complex numbers with an imaginary part (not imaginary in the sense of "existing only in the realms of the imagination"). Although the Fundamental Theorem guarantees those roots are there, it doesn't actually help you find them. To do that, you'll need to use known methods, like synthetic division, supplemented by some additional techniques.

### Descartes' Rule of Signs

Since the Fundamental Theorem can guarantee only the existence of roots, you need another tool to help you decipher how many of those guaranteed roots are positive roots and how many are negative. This is accomplished through **Descartes' Rule of Signs**, a method of divining between the two kinds of roots based on the number of sign changes between terms of a given polynomial. This is best illustrated with an example.

**Example 12:** Predict how many positive and negative real roots the polynomial function $f(x) = -1 + 3x + 2x^3 - x^2$ has, based on Descartes' Rule of Signs.

Arrange the terms of the polynomial in order of exponent, from the term with the highest exponent to the term with the lowest:

$$f(x) = 2x^3 - x^2 + 3x - 1$$

Ignore the coefficients of the terms, and pay attention only to their signs. Count the number of times the sign changes in adjacent terms as you go from left to right. Figure 3-1 demonstrates how to count the sign changes.

**Figure 3-1** As you progress from left to right, the terms in the function $f(x)$ change signs three times.

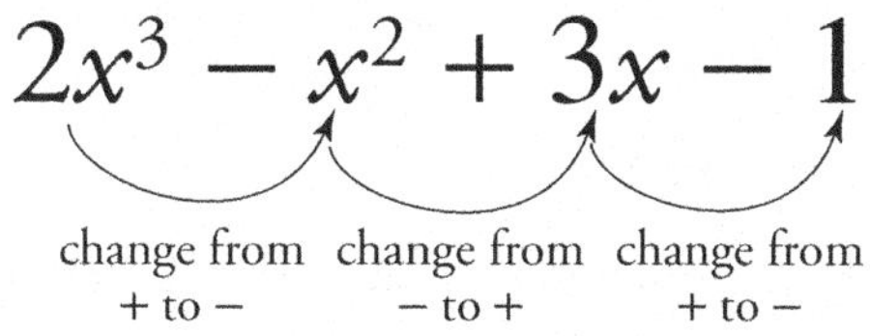


According to Descartes' Rule of signs, you will have either 3 positive real roots or 1 positive real root. To arrive at this conclusion, take the number of sign changes and continue to subtract 2, until your result is negative. For example, if there had been 6 sign changes in $f$, then Descartes' Rule of Signs tells you that $f$ has either 6, 4, 2, or 0 positive real roots. You do not know which is true.

To determine the possible number of negative real roots, begin by finding $f(-x)$:

$$f(-x) = 2(-x)^3 - (-x)^2 + 3(-x) - 1$$
$$f(-x) = -2x^3 - x^2 - 3x - 1$$

Count the number of sign changes in adjacent terms from left to right. This time, there are no sign changes at all. Therefore, $f$ has no negative real roots. If the number had been larger, say for example 5, you would have had to consider the possibility of 5, 3, or 1 negative real root(s), again following the procedure of subtracting 2 as you did when counting positive roots.

## The Rational Root Test

In Example 11, you were only able to factor the function if first given one of its roots. Without that initial root, you cannot perform synthetic division and determine what the other roots are.

Therefore, it is critical to apply a test whose purpose is to help you locate one or more roots of a function so that you are able to use the factor theorem and completely factor the function at hand. The **Rational Root Test** (also called the *Rational Zero Test*) does exactly that. It provides a list of candidates that *may or may not* be roots of a function, based on that function's leading coefficient and constant term.

The Rational Root Test says that a polynomial with leading coefficient $a$ and constant term $b$ can possess rational roots only of the form $\pm\frac{p}{q}$, where $p$ is a factor of $b$ and $q$ is a factor of $a$. It cannot discern which are actually roots; that task falls to you and synthetic division.

**Example 13:** Find all possible rational roots of

$$g(x) = 4x^3 - 2x^2 + 13x - 5$$

List the factors of the constant and leading coefficient, ignoring the signs of those terms:

$$\text{Factors of 5: } 1, 5$$

$$\text{Factors of 4: } 1, 2, 4$$

The list of possible rational roots of $g(x)$ will be the list of all possible combinations of the constant's factors divided by the leading coefficient's factors. (Additionally, each could be positive or negative, and that is indicated with a $\pm$ sign.)

$$\pm\frac{5}{4}, \pm\frac{5}{2}, \pm 5, \pm\frac{1}{4}, \pm\frac{1}{2}, \pm 1$$

Because there are 6 possible combinations, each of which could be positive or negative, there are 12 possible rational roots for $g(x)$.

### Determining roots

Using the Rational Root Test in conjunction with Descartes' Rule of Signs, you are now able to factor (and hence find the roots of) a far greater number of polynomials.

**Example 14:** Find all roots of the following polynomials.

(a) $f(x) = 3x^3 - 2x^2 - 7x - 2$

According to the Rational Root Test, these are the potential rational roots:

$$\pm\frac{1}{3}, \pm\frac{2}{3}, \pm 1, \pm 2$$

Try synthetic division with root candidates until you find one with a 0 remainder. In this case, 2 works. (If the instructor allows you, use a graphing utility to graph the function, and test any roots generated by the Rational Root Test that appear to be $x$-intercepts of the graph.)

$$\begin{array}{r|rrrr} 2 & 3 & -2 & -7 & -2 \\ & & 6 & 8 & 2 \\ \hline & 3 & 4 & 1 & 0 \end{array}$$

According to Descartes' Rule of Signs, there is only one positive root, so it must be $x = 2$. Find the other roots by factoring:

$$(x-2)(3x^2 + 4x + 1)$$

$$(x-2)(3x+1)(x+1)$$

Set each factor equal to 0 and solve; the roots are $x = -1, -\frac{1}{3}$, and 2.

(b) $g(x) = 2x^4 + 9x^3 - 21x^2 - 77x + 15$

The Rational Root Test lists these as possible roots:

$$\pm\frac{1}{2}, \pm 1, \pm\frac{3}{2}, \pm\frac{5}{2}, \pm 3, \pm 5, \pm\frac{15}{2}, \pm 15$$

Notice that 3 is a root:

$$\begin{array}{r|rrrrr} 3 & 2 & 9 & -21 & -77 & 15 \\ & & 6 & 45 & 72 & -15 \\ \hline & 2 & 15 & 24 & -5 & 0 \end{array}$$

You currently have $g(x)$ factored as

$$(x-3)(2x^3 + 15x^2 + 24x - 5)$$

You still need to find another rational root, but when testing now, perform synthetic division on the remaining cubic, rather than the original quartic polynomial. The only other rational root is –5:

$$\begin{array}{r|rrrr} -5 & 2 & 15 & 24 & -5 \\ & & -10 & -25 & 5 \\ \hline & 2 & 5 & -1 & 0 \end{array}$$

Thus, $g(x) = (x - 3)(x + 5)(2x^2 + 5x - 1)$. The quadratic cannot be factored, so you have to use the quadratic equation to find the final two roots. (They cannot be found using synthetic division, nor were they listed by the Rational Root Test, because they are irrational.)

$$x = \frac{-5 \pm \sqrt{5^2 - 4(2)(-1)}}{2(2)} = \frac{-5 \pm \sqrt{33}}{4}$$

The four roots are

$$x = -5, 3, \frac{-5 + \sqrt{33}}{4}, \text{ and } \frac{-5 - \sqrt{33}}{4}$$

(c) $h(x) = x^4 + 20x^2 + 64$

Factor $h(x)$ as you would a quadratic:

$$h(x) = (x^2 + 4)(x^2 + 16)$$

Set each factor equal to 0 and solve:

$$x^2 = -4 \text{ or } x^2 = -16$$

$$x = \sqrt{-4} \text{ or } x = \pm\sqrt{-16}$$

$$x = \pm 2i \text{ or } x = \pm 4i$$

All four roots are imaginary: $-4i$, $-2i$, $2i$, and $4i$. Note that Descartes' Rule of Signs warns you that this function will have neither positive nor negative real roots, so all 4 roots must be imaginary.

## Advanced Graphing Techniques

There are two more graph visualization methods you should master before this study in polynomial functions comes to a close. The first will help you determine what the ends of a polynomial graph will do, and the second will assist you in drawing *rational functions*, the quotient of two polynomials.

### The Leading Coefficient Test

Once you are capable of calculating the intercepts and roots of a graph, you can generally get a good idea of the graph's shape. You won't be able to construct more exact graphs without plotting point after point until you understand more advanced calculus concepts, but in the meantime, paying attention to the **end behavior** of a graph can help you visualize any polynomial function.

The *end behavior* of a graph is a description of the direction a graph is heading as its $x$-values get either infinitely positive or infinitely negative. In other words, the *end behavior* describes what direction the graph is heading at the far right and left edges of the coordinate axes.

Some end behavior can be determined just by examining the degree of the polynomial:

- Polynomial functions of even degree have the same right- and left-hand end behavior. In other words, the ends of the graph either both go up or both go down. (Consider the graph of $y = x^2$; both ends of the parabola go up.)
- Polynomial functions of odd degree have opposing right- and left-hand end behavior. For example, if the right end of the graph goes up, the left end goes down. (Consider the graph of $y = x^3$; the graph goes down to the left, but up to the right.)

The Leading Coefficient Test sheds additional light on the end behavior of polynomials, and only requires you to know the leading coefficient of the polynomial in question:

- If the leading coefficient of a polynomial with an even degree is positive, both the left and right ends will go up. If the leading coefficient is negative, both ends will go down.
- If the leading coefficient of a polynomial with an odd degree is positive, the left end will go down and the right end will go up. The opposite is true if the leading coefficient is negative.

**Example 15:** Describe the end behavior of each polynomial.

(a) $f(x) = -2x^6 - 3x^5 + 4x - 19$

Since $f$ has an even degree and a negative leading coefficient, the graph will go down to the left and down to the right.

(b) $g(x) = -2 - 7x - 8x^2 + 4x^3$

Even though $g$ is not written in the correct exponential order, the leading coefficient is positive (4) and the degree is odd (3), so $g$ will go down to the left and up to the right.

## Finding rational asymptotes

Remember, a **rational function** is a function defined as the quotient of two polynomials. Most rational function graphs contain **asymptotes**, lines that represent values that the function cannot attain. Asymptotes are usually drawn as dotted lines on a graph so that they are not confused as part of the function itself. Visually speaking, a graph is shaped by *asymptotes* because it bends to avoid making contact with the lines.

There are three basic types of asymptotes. The following list describes how to find each kind of asymptote for the generic rational function $f(x) = \frac{n(x)}{d(x)}$.

- **Vertical asymptote:** Any value $a$ for which $d(a) = 0$ but $n(a) \neq 0$, has a corresponding vertical asymptote with equation $x = a$.
- **Horizontal asymptote:** Compare the degrees of $n(x)$ and $d(x)$.

  If the degree of $n$ is greater than the degree of $d$, then $f$ has no horizontal asymptotes.

  If the degree of $d$ is greater than the degree of $n$, then $f$ has only one horizontal asymptote: $y = 0$.

  If the degrees of $n$ and $d$ are equal, then $f$ has the horizontal asymptote $y = \frac{a}{b}$, where $a$ is the leading coefficient of $n(x)$ and $b$ is the leading coefficient of $d(x)$.
- **Slant asymptote**: A slant asymptote (also called an *oblique asymptote*) is a linear asymptote that is neither vertical nor horizontal. The function $f(x)$ has a *slant asymptote* only if the degree of $n(x)$ is one greater than the degree of $d(x)$. To find its equation, divide $n(x)$ by $d(x)$ to get $q(x) + r(x)$, where $q(x)$ is the quotient and $r(x)$ is the remainder. The slant asymptote will have equation $y = q(x)$.

**Example 16:** Provide the equations for all asymptotes of the function $f(x) = \frac{x^3 + x^2 - 14x - 24}{x^2 - 9x - 22}$.

Factor the numerator and denominator of $f(x)$:

$$f(x) = \frac{(x-4)(x+2)(x+3)}{(x-11)(x+2)}$$

The $x$-values 11 and $-2$ make the denominator 0, but $f$ will only have the vertical asymptote $x = 11$, since $-2$ makes the numerator equal to 0 as well.

Because the degree of the numerator is greater than the denominator, $f$ will have no horizontal asymptotes, but since it is exactly one greater, $f$ will have a slant asymptote. To find it, perform long division on $f$:

$$f(x) = x + 10 + \frac{98x + 196}{x^2 - 9x - 22}$$

The slant asymptote for $f$ has the equation $y = x + 10$.

To graph a rational equation, plot the intercepts, draw the asymptotes, and then substitute a sufficient number of $x$-values into the function to get a good idea of the graph's shape.

**Example 17:** Graph the rational function $g(x) = \frac{x^2 - x - 12}{x^2 - 9}$.

Factor the numerator and denominator of $g(x)$:

$$g(x) = \frac{(x-4)(x+3)}{(x-3)(x+3)}$$

The function has $x$-intercept 4 and $y$-intercept $\frac{4}{3}$. The equations of the asymptotes are $x = 3$ and $y = 1$. (Since the numerator and denominator have the same degree, the horizontal asymptote comes from the quotient of the leading coefficients of the fraction, both of which are 1.) Because $x = -3$ makes both the numerator and denominator 0, the graph will have a hole at that $x$-value, as illustrated in Figure 3-2.

**Figure 3-2** The graph of $g(x)$, the answer to Example 17.

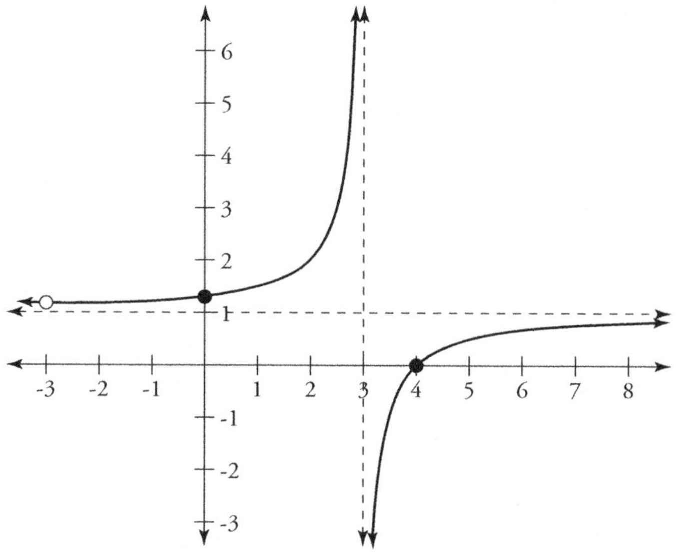

## Chapter Checkout

### Q&A

1. Factor the expressions completely.
   (a) $x^3y + x^2z - 4xy - 4z$
   (b) $x^3 - 64y^3$
2. List all possible rational roots of $h(x) = 2x^3 + 5x^2 - 11x + 6$.
3. Find all roots of each function.
   (a) $f(x) = 8x^4 + 22x^3 - 41x^2 - x + 12$
   (b) $g(x) = 3x^3 - 22x^2 + 36x - 5$
4. Identify the asymptotes for the graph of

$$g(x) = \frac{x+1}{3x^2 + 7x - 20}$$

**Answers: 1. (a)** $(xy + z)(x + 2)(x - 2)$ **(b)** $(x - 4y)(x^2 + 4xy + 16y^2)$
**2.** $\pm\frac{1}{2}, \pm 1, \pm\frac{3}{2}, \pm 2, \pm 3, \pm 6$ **3. (a)** $-4, -\frac{1}{2}, \frac{3}{4}, 1$ **(b)** $5, \frac{7+\sqrt{37}}{6}, \frac{7-\sqrt{37}}{6}$
**4.** $x = \frac{5}{3}, x = -4, y = 0.$

# Chapter 4

# EXPONENTIAL AND LOGARITHMIC FUNCTIONS

## Chapter Check-In

- ❑ Graphing exponential and logarithmic functions
- ❑ Calculating logs of any base
- ❑ Applying logarithmic properties
- ❑ Solving exponential and logarithmic equations
- ❑ Understanding word problems containing exponential functions

Functions containing exponents are nothing new or unique. In fact, in Chapter 3, you worked extensively with polynomial functions, which contain plenty of exponents. The difference between polynomial functions and exponential functions is that polynomial functions contain variables raised to numerical powers, and exponential functions contain numbers raised to variable powers.

In this chapter, you'll evaluate, graph, and solve exponential equations and functions. You'll also be introduced to the logarithm, which serves as the exponential function's inverse.

## Exponential Functions

An **exponential function** is of the form $f(x) = a^x$, for some real number $a$, as long as $a > 0$. While exponential functions accept any real number input for $x$, the range is limited to positive numbers.

## Natural exponential function

Although you will deal with many, the most common exponential function you'll encounter is the **natural exponential function**, written as $f(x) = e^x$. Although the base $e$ looks just as generic as the base $a$ in our definition of *exponential function*, it is not. The $e$ stands for **Euler's number**, and represents a standard, commonly known, irrational constant, sort of like $\pi$. (Note that "Euler" is pronounced "OIL-er," not "YOU-ler.")

$$e \approx 2.718281828459045 23...$$

Although the decimal digits in $e$ appear to repeat themselves at first (2.718281828...), they soon diverge into an non-repeating and non-terminating pattern.

You are not expected to memorize $e$, just as you are not expected to memorize $\pi$; therefore, answers can be left in terms of $e$ (such as $12e^5$) and are still considered simplified. If you are expected to evaluate the natural exponential function, you will be allowed to use a calculator.

$$12e^5 = 12(148.413159...) \approx 1780.958$$

All scientific and graphing calculators have an $e$ button, but be aware that in some tools and graphing software packages, the $e^x$ button is labeled as "exp."

## Graphs of exponential functions

In Chapter 2, you were presented with the graph of the natural exponential function, $y = e^x$. All exponential functions have the same basic shape and properties; they only differ in steepness, according to the constant raised to the $x$ power. The larger the constant, the more quickly and steeply the graph will rise once $x > 0$.

Consider the graph of $f(x) = 2^x$ in Figure 4-1, plotted by substituting a small collection of integers into $f$.

From the table, it is clear why the graph of $f$ gets closer and closer to the $x$-axis as $x$ gets more and more negative. Since a negative input becomes a negative exponent, your result will be a fraction. The larger the negative input, the smaller the value of the fraction. However, once the inputs become positive, the graph grows quickly.

$$f(1) = 2, f(2) = 4, f(3) = 8, f(4) = 16, f(5) = 32, f(6) = 64, ...$$

**Figure 4-1** The graph of $f(x) = 2^x$.

| $x$ | $f(x) = 2^x$ | $f(x)$ |
|---|---|---|
| $-4$ | $2^{-4}$ | $\frac{1}{16}$ |
| $-3$ | $2^{-3}$ | $\frac{1}{8}$ |
| $-2$ | $2^{-2}$ | $\frac{1}{4}$ |
| $-1$ | $2^{-1}$ | $\frac{1}{2}$ |
| 0 | $2^0$ | 1 |
| 1 | $2^1$ | 2 |

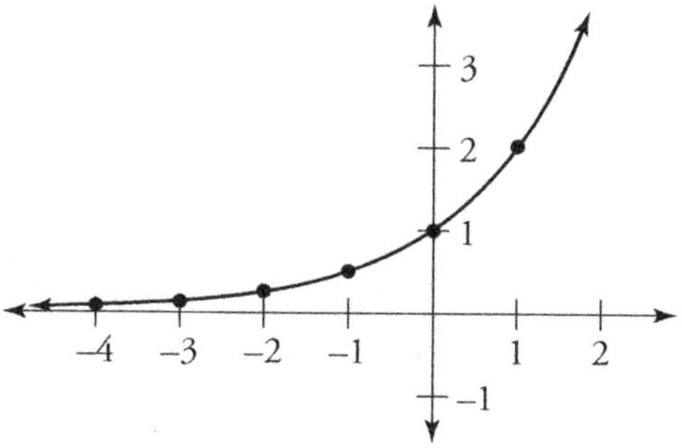


One other thing becomes clear upon examination of $f(x)$. Note that $f(0) = 1$. That is not true only when the base of the exponential function is 2. In fact, no matter what the base $a$ in $g(x) = a^x$, the graph of $g$ will contain the point (0,1) since any positive number $a$ raised to the zero power will be 1, according to exponential rule six from Chapter 1.

Since you can be sure the point (0,1) will appear in an exponential graph, use it as the anchor point for your sketch if asked to transform the graph of an exponential function.

**Example 1:** Sketch the graph of $h(x) = -3^{x+2} - 1$.

Approach this problem like the graph transformations in Chapter 2. Your graph should be $y = 3^x$ reflected across the $x$-axis, moved two units to the left, and one unit down. Think about this in terms of the anchor point (0,1). The initial reflection changes the point to (0,–1); once shifted left and down, the anchor point ends up at (–2,–2). The former horizontal asymptote of $y = 0$ has also moved down one, so the new horizontal asymptote is $y = -1$. See Figure 4-2.

**Figure 4-2** The graph of $y = 3^x$, shown as a dotted curve, and the graph of $h(x)$, the answer to Example 1.

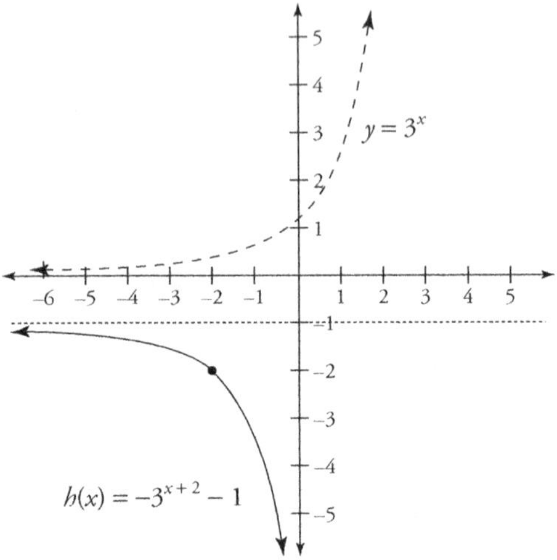


## Logarithmic Functions

Whereas an exponential function answers the question "A number raised to a power equals what?" a **logarithmic function** (or *log function*) answers the question "To what power must I raise a number to get another number?" In other words, the output for a *logarithmic function* is in actuality an exponent.

Specifically, the logarithmic expression $\log_c x$ (read "the log base $c$ of $x$") asks the question: $c$ to what power equals $x$? Thus, the equations $\log_c x = n$ and $c^n = x$ mean precisely the same thing.

**Example 2:** Find $x$ in each of the equations.

(a) $\log_3 81 = x$

This expression is the equivalent of $3^x = 81$, so $x = 4$. It answers the question "3 to what power equals 81?"

(b) $\log_2 x = -5$

Rewrite as the equation $2^{-5} = x$ and evaluate; $x = \frac{1}{32}$.

(c) $\log_x 125 = 3$

Rewrite as the equation $x^3 = 125$ and take the cube root of each side; $x = 5$.

(d) $\log_a 1 = x$, where $a$ is a positive integer

Rewrite as the equation $a^x = 1$. No matter the value of $a$, only one $x$ value will result in a value of 1: $x = 0$, since any positive number raised to the 0 power is 1.

### Natural and common logs

Although a logarithm's base can be any positive number (except for 1, since 1 raised to any real number will still be 1), there are two bases you'll encounter most often.

- **Base 10.** A logarithm of base 10 is called a **common log**. In fact, if a logarithmic expression is written without specifying a base, that base is understood to be 10, in the same way that an unwritten exponent is understood to be 1.

$$\log 1000 = 3, \text{ since } 10^3 = 1000$$

- **Base *e*.** Just like the exponential function with base $e$ is called the *natural exponential function*, the logarithm with base $e$ is called the **natural logarithm**. It is used so frequently that it has its own notation: $\ln x$, and is read "the natural log of $x$" or "L-N of $x$," in which case you actually say the letters L and N. Therefore, $\ln x$ is the same thing as $\log_e x$.

### Inverse relationship

Since exponential and logarithmic functions of the same base are inverses of one another, if you compose the two functions together, they will cancel one another out.

$$\log_a\left(a^x\right) = a^{\log_a x} = x$$

Since you will see common and natural logs most often, here is that inverse relationship expressed in terms of their respective bases:

$$\log\left(10^x\right) = 10^{\log x} = x$$

$$\ln\left(e^x\right) = e^{\ln x} = x$$

### Graphs of logarithmic functions

Since logarithmic and exponential functions are one another's inverses, it is easy to construct the graph of any logarithmic function $y = \log_a x$ based on the corresponding graph of $y = a^x$. In Chapter 2 you learned that graphs of inverse functions are reflections of one another across the line $y = x$, since each graph contains the coordinates of the other graph, with each coordinate pair reversed. It is no surprise, then, that because all exponential graphs of the form $y = a^x$ contain the point (0,1), then all logarithmic graphs of the form $y = \log_a x$ contain the point (1,0).

In Figure 4-3, you can visually verify that the graphs of the natural logarithmic and natural exponential functions (both of which you already memorized in Chapter 2) are, indeed, reflections of one another about the line $y = x$.

**Figure 4-3** The graphs of $y = e^x$ and $y = \ln x$ are reflections of one another about the line $y = x$, as are all inverse functions.

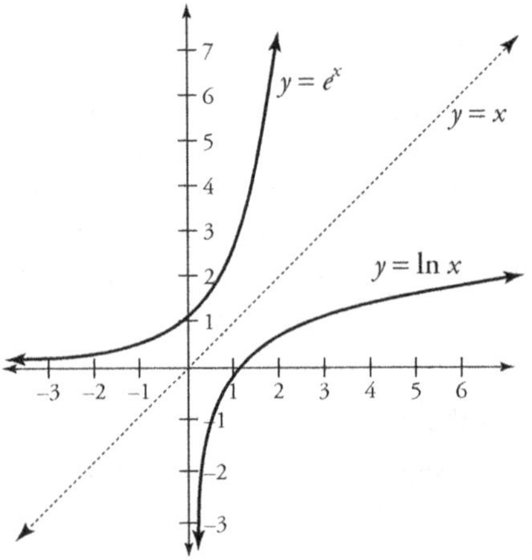


Note that the domain of $\ln x$, like all logarithmic functions of form $y = \log_a x$, is $(0,\infty)$. Although it might appear that the $y$ values of the logarithmic graph "level out," as if approaching a horizontal asymptote, they do not. In fact, a logarithmic graph will grow infinitely tall, albeit much, much slower than its sister the exponential function. A range of $(-\infty,\infty)$ for the logarithmic functions makes sense, since their inverses are exponential functions and have domains of $(-\infty,\infty)$.

In order to graph transformations of the logarithmic function (such as reflections, shifts, and stretches), follow the procedures outlined in Chapter 2, just as you did when graphing exponential transformations earlier in this chapter.

### Change of base formula

With the aid of a scientific or graphing calculator, it is a simple matter to evaluate a logarithm. (It is not appropriate or necessary to learn to calculate complex decimal values of logarithms by hand.) However, you may notice that most computational tools have only two logarithmic buttons: one for common log and one for natural log. Thus, while it may be simple to calculate these values:

$$\log 6 \approx .77815$$

$$\ln 49 \approx 3.89182$$

you'll need to use the **change of base formula** to calculate the values of logs whose base is neither 10 nor $e$.

According to this formula, you can rewrite a logarithm of base $c$ as a quotient of two logs with a different base, $n$.

$$\log_c x = \frac{\log_n x}{\log_n c}$$

Even though you can choose any base $n$, you should pick either 10 or $e$, since that will allow you to use a calculator to find its decimal value.

**Example 3:** Evaluate $\log_5 9$ using a calculator.

Rewrite the logarithm as a quotient of natural logs by means of the change of base formula.

$$\begin{aligned}\log_5 9 &= \frac{\ln 9}{\ln 5} \\ &= \frac{2.197224577...}{1.609437912...} \\ &\approx 1.3652\end{aligned}$$

You could also have rewritten $\log_5 9$ as $\frac{\log 9}{\log 5}$, and the final result would have been the exact same decimal value.

## Properties of Logarithms

There are three major logarithmic properties which allow you to both expand and compress logarithmic expressions. The rules are listed in terms of common logs, but the properties hold true for all valid logarithmic bases.

- **Log property 1: $\log (xy) = \log x + \log y$**

  In other words, the log of a product is equal to the sum of the logs of its factors.

  Though rigorously proving this logarithmic property is unnecessary, you can easily see why it's true using a real number example and a calculator to evaluate the expressions therein.

$$\ln 15 = \ln (3 \cdot 5) = \ln 3 + \ln 5$$
$$2.708050201... = 1.098612289... + 1.609437912\ ...$$
$$2.708050201... = 2.708050201...$$

- **Log property 2: $\log x^a = a \log x$**

  An exponent within a logarithm can be "drawn out" of the logarithm and placed in front of it as a coefficient. This is due to the fact that an exponent indicates repeated multiplication and, according to logarithmic properties, multiplication within one log can be rewritten as the sum of separate logs.

  Consider $\ln x^4$. Rewrite the logarithm without using an exponent.

$$\ln (x \cdot x \cdot x \cdot x)$$

  Rewrite the string of $x$'s using log property 1.

$$\ln x + \ln x + \ln x + \ln x$$

  Since these are all like terms, you can simplify that expression.

$$4\ln x$$

  Even though this is only an example, and not a rigorous proof, it sheds light on why, exactly $\ln x^4 = 4\ln x$, and (by extension) why log property 2 works.

- **Log property 3: $\log \frac{x}{y} = \log x - \log y$**

  In other words, the log of a quotient is equal to the difference of the individual logs. It is easy to prove this property formally.

  Consider the logarithm $\log \frac{x}{y}$. The quotient can be rewritten as a product, if you use a negative exponent.

$$\log (x \cdot y^{-1})$$

Rewrite this expression using log property 1.

$$\log x + \log y^{-1}$$

Apply log property 2 to the second term.

$$\log x + (-1)\log y$$
$$\log x - \log y$$

Therefore, $\log \frac{x}{y} = \log x - \log y$.

**Example 4:** Fully expand each of the following expressions by applying logarithmic properties, and simplify the result if necessary.

(a) $\log 2x^2y^3$

Apply log property 1.

$$\log 2 + \log x^2 + \log y^3$$

Finish by applying log property 3.

$$\log 2 + 2\log x + 3\log y$$

(b) $\log_2 \frac{\sqrt[3]{x+1}}{2x}$

Begin by applying log property 3.

$$\log_2 \sqrt[3]{x+1} - \log_2 2x$$

The first term requires log property 2, and the other property 1.

$$\frac{1}{3}\log_2 (x+1) - (\log_2 2 + \log_2 x)$$

Remember, since $\log_a a^n = n$ (because exponential and logarithmic functions with the same base are inverses and cancel one another out when composed together), then $\log_2 2^1 = 1$, so simplify.

$$\frac{1}{3}\log_2 (x+1) - \log_2 x - 1$$

**Example 5:** Rewrite each of the following expressions as a single log.

(a) $3\ln x - 2\ln (x+1)$

Apply log property 2.

$$\ln x^3 - \ln (x+1)^2$$

Now apply log property 3.

$$\ln \frac{x^3}{(x+1)^2}$$

(b) $\log x - 2\log y - \log z + 3\log w$

Begin with log property 2. Every positive term belongs in the numerator, and every negative term in the denominator.

$$\log \frac{xw^3}{y^2 z}$$

## Solving Exponential and Logarithmic Equations

Take advantage of the fact that exponential and logarithmic equations cancel one another out to solve equations containing those elements. Remember, $\log_a (a^x) = x$ and $a^{\log_a x} = x$.

### Exponential equations

Your ultimate goal when solving exponential equations is to isolate the term with the exponent, $a^x$, on one side of the equation and cancel it out by taking $\log_a$ of both sides of that equation. This leaves the form "$x =$" and gives you the solution.

**Example 6:** Solve the equations.

(a) $3^x = 243$

Both sides of this equation can be written with base 3.

$$3^x = 3^5$$

Take $\log_3$ of both sides of the equation.

$$\log_3 3^x = \log_3 3^5$$
$$x = 5$$

(b) $5^x = 13$

Since 13 cannot be rewritten as an exponent with base 5, proceed to logging both sides of the equation.

$$\log_5 5^x = \log_5 13$$
$$x = \log_5 13$$

Use the change of base formula.

$$\log_5 13 = \frac{\ln 13}{\ln 5} \approx 1.59369$$

(c) $2e^{3x} - 3 = 15$

Isolate $e^{3x}$ on the left side of the equation.

$$2e^{3x} = 18$$
$$e^{3x} = 9$$

Take the natural log of both sides of the equation.

$$\ln (e^{3x}) = \ln 9$$
$$3x = \ln 9$$
$$x = \frac{\ln 9}{3} \approx .73241$$

(d) $e^{2x} - 2e^x - 15 = 0$

Notice that $e^{2x} = (e^x)^2$, according to exponential rule 3 from Chapter 1. Therefore, this is really a quadratic equation and can be solved by factoring.

$$(e^x)^2 - 2e^x - 15 = 0$$
$$(e^x - 5)(e^x + 3) = 0$$
$$e^x = 5 \quad \text{or} \quad e^x = -3$$
$$\ln e^x = \ln 5 \quad \text{or} \quad \ln e^x = \ln -3$$
$$x = \ln 5 \text{ or } \ln -3$$

Only one of those solutions is valid. Since the domain of $\ln x$ is $(0,\infty)$, you must discard the solution $x = \ln -3$.

## Logarithmic equations

Your ultimate goal when solving logarithmic equations is to isolate the logarithmic function $\log_a x$ on the left side of the equation. To cancel it out, you'll raise $a$ to the power of each side of the equation in a process called **exponentiating.** In other words, to solve the equation

$$\log_a x = c$$

introduce exponential functions to both sides like this:

$$a^{\log_a x} = a^c$$
$$x = a^c$$

**Example 7:** Solve the equations.

(a) $\log_7 x = 2$

Although you can solve this intuitively, use it as an opportunity to practice exponentiating. Since the log has base 7, rewrite each side of the equation as an exponent of 7.

$$7^{\log_7 x} = 7^2$$

$$x = 49$$

(b) $\ln x - \ln (x + 1) = \ln 3$

Subtract ln 3 from both sides and compress all of the logs into a single log, using log properties.

$$\ln x - \ln (x + 1) - \ln 3 = 0$$

$$\ln \frac{x}{3(x+1)} = 0$$

Exponentiate both sides with a base of $e$ to cancel the ln.

$$e^{\ln \frac{x}{3(x+1)}} = e^0$$

$$\frac{x}{3x+3} = 1$$

$$3x + 3 = x$$

$$x = -\frac{3}{2}$$

Notice that the solution cannot be substituted into the original equation. The expression $\ln (x + 1)$ will become invalid because the natural log function cannot accept a negative input. Therefore, there is no solution to this equation.

(c) $\log 3x = \frac{1}{4}$

Since this is the common log, Exponentiate with base 10.

$$10^{\log 3x} = 10^{1/4}$$

$$3x = \sqrt[4]{10}$$

$$x = \frac{\sqrt[4]{10}}{3} \approx .5928$$

(d) $(\ln x)^2 - 2\ln (x^4) = 20$

Rewrite the second term using log property 2.

$$(\ln x)^2 - 4 \cdot 2\ln x = 20$$

$$(\ln x)^2 - 8\ln x - 20 = 0$$

This quadratic equation can be solved by factoring.

$$(\ln x - 10)(\ln x + 2) = 0$$

$$\ln x - 10 = 0 \quad \text{or} \quad \ln x + 2 = 0$$

$$\ln x = 10 \quad \text{or} \quad \ln x = -2$$

$$x = e^{10}, e^{-2}$$

Simplified solutions do not contain negative exponents, so the final answers are $x = \frac{1}{e^2}$ and $e^{10}$. These solutions both prove valid if substituted back into the original equation.

## Exponential Word Problems

The two main types of word problems usually associated with the study of exponential functions involve either compound interest or exponential growth and decay.

### Compound interest

The more often interest is compounded upon an initial investment (also called the **principal**), the more interest will accrue. Thus, monthly compounding is preferable to annual or quarterly compounding. However, the best option is continuous compounding. Interest compounded a finite number of times per year is calculated using a different method than continuous compounding.

- **Interest compounded *n* times per year**

  A principal, $P$, which is compounded $n$ times per year at an annual interest rate $r$ (expressed as a decimal) will have balance $B(t)$ after a period of $t$ years:

  $$B(t) = P\left(1 + \frac{r}{n}\right)^{nt}$$

- **Interest compounded continuously**

  A principal, $P$, compounded continuously at an annual interest rate $r$ (expressed as a decimal) will have balance $B(t)$ after a period of $t$ years:

  $$B(t) = Pe^{rt}$$

**Example 8:** How much faster will an initial investment of $15,000 grow to a balance of $20,000 if the 4.5% annual interest rate is compounded continuously rather than monthly?

You need to calculate two kinds of interest, but in each you know that $P = 15{,}000$, $B(t) = 20{,}000$, and $r = .045$. Plug what you know into the formula for continuous compounding of interest.

$$20{,}000 = 15{,}000e^{(.045)t}$$
$$\frac{4}{3} = e^{.045t}$$

To solve for $t$, take the natural log of both sides.

$$\ln\frac{4}{3} = .045t$$
$$t = \frac{\ln\frac{4}{3}}{.045} \approx 6.39293 \text{ years}$$

Now calculate the time it will take when the interest is compounded $n = 12$ times per year using the other interest formula.

$$20,000 = 15,000\left(1 + \frac{.045}{12}\right)^{12t}$$
$$\frac{4}{3} = (1.00375)^{12t}$$

To solve this, take the log base 1.00375 of both sides of the equation.

$$\log_{1.00375}\frac{4}{3} \approx 12t$$

Calculate the left side using the change of base formula; solve for $t$.

$$t = \frac{76.8589706242}{12} \approx 6.40491 \text{ years}$$

Therefore, continuous interest brings you to the new balance $6.40491 - 6.39293 = .00198$ years (or $.00198 \times 365 \approx 4.373$ days) faster.

### Growth and decay

If a quantity either increases or decreases at a rate which is proportional to the size of the quantity itself, it exhibits either *exponential growth* or *exponential decay*, respectively. Mathematically speaking, the size of the quantity is

$$P(t) = Ne^{kt}$$

where $N$ is the original quantity, $t$ is the amount of time elapsed, and $k$ is a constant of proportionality. Note that this formula is a clone of the formula for continuously compounding interest, only expressed with different variables.

You will rarely if ever be given the value $k$ explicitly; it is not as easy to pick out from the given information as were the variables for compound interest. In fact, it is usually the first order of business in an exponential change problem to calculate what $k$ is based on the starting and ending sizes of the quantity in question and the amount of time that has elapsed.

The most common exponential growth and decay problems involve radioactive half-life (the amount of time it takes the mass of a given radioactive element to decrease by exactly half) and bacteria growth.

**Example 9:** One of the radioactive isotopes of the element Californium, Cu-250, has a half-life of 13 years. How long will it take 500 grams of Cu-250 to decay to 75 grams?

Your first task is to calculate $k$. Since Cu-250 has a half-life of 13 years, you know that once $t = 13$ years have elapsed, the original $N = 500$ grams will decrease to exactly half, $P(13) = 250$ grams. Plug in the values of the known variables to solve for $k$.

$$250 = 500e^{k(13)}$$

$$\frac{1}{2} = e^{13k}$$

$$\ln\frac{1}{2} = 13k$$

$$k \approx -.053319013889$$

The more decimal places you use for $k$, the more accurate your answer will be. Now that you know $k$, you can use it to solve the actual question posed by the problem. This time, $P(t) = 75$, and you are solving for $t$.

$$75 = 500e^{(-.053319013889)t}$$

$$.15 = e^{(-.053319013889)t}$$

$$t = \frac{-\ln 0.15}{0.053319013889} \approx 35.581 \text{ years}$$

## Chapter Checkout

### Q&A

**1.** Solve each equation.

(a) $\log \frac{1}{100} = x$

(b) $\log_x 5 = 2$

(c) $\log_8 x = \frac{4}{3}$

(d) $\ln e^{17} = x$

**2.** Evaluate $\log_2 15$ using a calculator.

**3.** Rewrite as a single logarithm:

$$4\ln (x + 3y) - 2\ln z + \frac{1}{2}\ln w$$

**4.** Solve each equation.

(a) $\ln 2x + \ln 5 = 3$

(b) $6^x = 216$

**5.** If a bacteria population triples every three hours, how long will it take a population of 20 colonies to grow to 1,000 colonies?

**Answers: 1. (a)** $-2$ **(b)** $\sqrt{5}$ **(c)** 16 **(d)** 17 **2.** 3.90689 **3.** $\ln \frac{(x+3y)^4 \cdot \sqrt{w}}{z^2}$ **4. (a)** $\frac{e^3}{10}$ **(b)** 3 **5.** 10.683 hours

# Chapter 5

# TRIGONOMETRY

## Chapter Check-In

- ❑ Working with angles in standard position
- ❑ Understanding basic trigonometric functions
- ❑ Defining unit circle values for common angles
- ❑ Graphing trigonometric functions
- ❑ Calculating the values of inverse trigonometric functions

At its core, trigonometry is the study of triangles, so everything you'll learn about trigonometry is either based upon, derived from, or applicable to triangles. In this chapter, you'll develop a set of functions based on a preliminary study of angles and their properties. Those familiar functions (including sine, cosine, and tangent) can help you uncover just about any attribute of any triangle, whether it be a side, angle measurement, or its area.

## Measuring Angles

Since angles are one of the basic components of triangles, it's important to establish a set of standard terms and common practices with regards to angles before you can begin to manipulate them. Furthermore, it's important to learn a new technique for communicating angle measurement, as few angles in precalculus and calculus are measured in degrees, the units with which you're probably the most familiar.

### Characteristics of angles

An **angle** is made up of two nonoverlapping rays that share the same endpoint; that endpoint is called the **vertex** of the angle. The *ray* marking the beginning of the angle is called the **initial side**, whereas the ray which marks the end of the angle is the **terminal side**. Any angle whose initial side lies on the positive $x$-axis and whose vertex lies on the origin of the coordinate plane is said to be in **standard position**. If the terminal side of the angle falls upon a coordinate axis, the angle is a described as a **quadrantal**.

Whereas variables representing real numbers are usually given names like $x$, $y$ and $z$, variables representing angles are usually labeled with Greek letter variables, such as $\theta$ (theta), $\alpha$ (alpha), $\beta$ (beta), and $\gamma$ (gamma).

In Figure 5-1, angle $\theta$ is created by ray $BC$ (the initial side) and ray $BA$ (the terminal side). Because its vertex, $B$, lies on the origin and ray $BC$ lies on the positive $x$-axis, $\theta$ is said to be in standard position.

**Figure 5-1** Angle θ is in standard position.

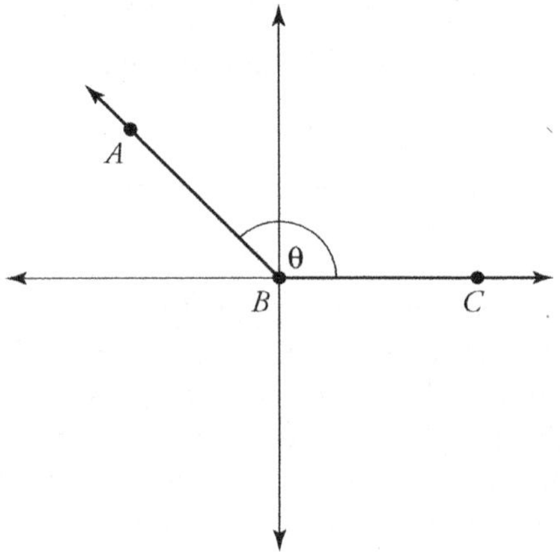


Since $\theta$ travels in a counterclockwise direction from its initial side to its terminal side, you know $\theta > 0$. On the other hand, angles that travel clockwise are negative.

## Degrees and radians

Students are usually most familiar with the angle measurement unit of *degrees* from geometry class. If an angle in standard position contains exactly one complete revolution about the origin, so that the terminal side overlaps the initial side, a **degree** is defined as 1/360th of that rotation. If you travel only one-fourth of that rotation, the angle formed is a right angle, and measures $\frac{1}{4} \cdot 360 = 90$ degrees, a result that should not be surprising. The notation used for degrees is a small circle written above and to the right of the angle measurement: $90°$.

Angle measurements using degrees can be expressed more exactly than simple whole degrees, just like the time of day can be expressed more exactly than in merely whole hours. In the same way you can say "We'll meet at 30 seconds after 5:45 pm," you can express fractions of a degree in **minutes** and **seconds**. In fact, degrees work similar to hours, because there are 60 *minutes* in one degree and 60 *seconds* in one *minute*. To write the degree measure "30 degrees, 12 minutes, 50 seconds," use this notation: $30^\circ 12' 50''$.

**Example 1:** Express the angle measurement $120^\circ 42' 37''$ as a degree in decimal form.

Express minutes and seconds as fractions of a degree. (Since there are 60 seconds in a minute, there are $60 \cdot 60 = 3600$ seconds in a degree.)

$$120 + \frac{42}{60} + \frac{37}{3600} \approx 120.710278^\circ$$

Because degree measurements get awkward once exact measurements involving minutes and seconds are involved, most problems in precalculus and calculus will use *radians*, another unit of angle measurement.

A **radian** is the measurement of an angle in standard position that, when extended to a circle of radius $r$ centered at the origin, will mark the endpoints of an arc whose length is also $r$. In Figure 5-2, the measure of angle $\theta$ is exactly 1 radian, because the arc it cuts out of the circle (or *intercepts*) is the same length as the radius of the circle. (Unlike degrees, there is no shorthand notation for radians. However, any angle containing "π" can be assumed in radian form, unless otherwise noted.)

**Figure 5-2** Because the darkened arc and the radius are both length $r$, $\theta$ measures exactly 1 radian.

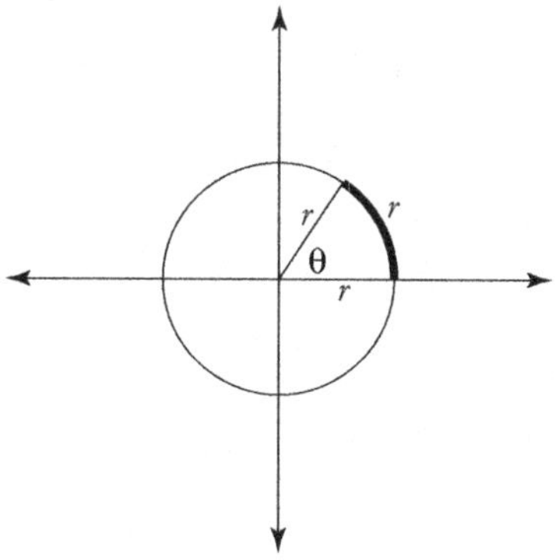

One radian measures just over 57.3°, and there are exactly $2\pi$ radians (approximately 6.283) in one full revolution around the origin (just like there are 360 degrees in one revolution). Therefore, the ratio

$$\frac{2\pi \text{ radians}}{360^\circ} = \frac{\pi \text{ radians}}{180^\circ}$$

makes it possible to convert from degrees to radians. Multiply any degree measurement by that ratio to translate degrees into radians. To change a measurement from radians to degrees, multiply by the reciprocal of the ratio.

**Example 2:** Complete the following conversions.

(a) Convert 210° to radians.

Multiply the degree measurement by the correct ratio and simplify.

$$210 \cdot \frac{\pi}{180} = \frac{210\pi}{180} = \frac{7\pi}{6} \text{ radians}$$

(b) Convert $-\frac{11\pi}{6}$ to degrees.

This time multiply by $\frac{180}{\pi}$ and simplify.

$$-\frac{11\pi}{6} \cdot \frac{180}{\pi} = -\frac{1980}{6} = -330^\circ$$

## Angle pairs

In geometry, you learned that two acute angles whose measurements add up to 90° are **complimentary**. This property holds true for angles expressed in radians as well, but the sum will need to equal the radian equivalent of 90°, which is $\frac{\pi}{2}$. Similarly, if the sum of two angle measurements totals 180° or $\pi$ radians, those angles are said to be **supplementary**.

**Example 3:** Find the indicated angles.

(a) The supplement of $\alpha = \frac{7\pi}{12}$

The sum of $\alpha$ and its supplement, $\theta$ will be $\pi$. Get common denominators and simplify.

$$\frac{7\pi}{12} + \theta = \pi$$

$$\theta = \frac{12\pi}{12} - \frac{7\pi}{12}$$

$$\theta = \frac{5\pi}{12}$$

(b) The complement of $\beta = \frac{2\pi}{9}$

The sum of $\beta$ and its complement, $\theta$, will be $\frac{\pi}{2}$. Again use common denominators

$$\frac{2\pi}{9} + \theta = \frac{\pi}{2}$$

$$\theta = \frac{\pi}{2} \cdot \frac{9}{9} - \frac{2\pi}{9} \cdot \frac{2}{2}$$

$$\theta = \frac{9\pi - 4\pi}{18}$$

$$\theta = \frac{5\pi}{18}$$

## Coterminal angles

Angles in standard position that share the same terminal ray are said to be **coterminal angles**. Consider Figure 5-3, in which three distinct angles share the same terminal side. Angle $\theta$ measures $\frac{2\pi}{3}$ radians (120°); it's the smallest positive angle possessing that terminal side. However, $\beta = -\frac{4\pi}{3}$ (–240°) also terminates at the exact same ray; the only difference is that $\beta$ approaches that terminal side by winding clockwise, rather than counterclockwise (since it is a negative angle.)

**Figure 5-3** Three coterminal angles whose common terminal ray lies in the second quadrant.

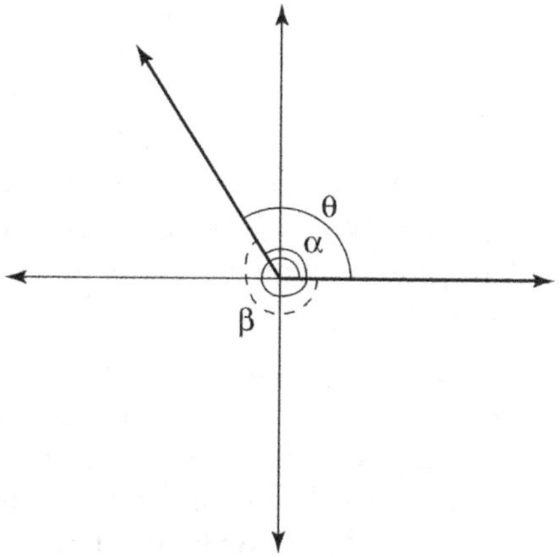

The other coterminal angle pictured, $\alpha$, is a positive angle like $\theta$, but winds completely around the origin once before terminating. To find its measure, add the radian equivalent of one full rotation ($2\pi$) to $\theta$.

$$\alpha = \theta + 2\pi$$
$$\alpha = \frac{2\pi}{3} + 2\pi$$
$$\alpha = \frac{8\pi}{3}$$

In fact, this is the method used to calculate the measures of coterminal angles. Simply add $2\pi$ or subtract $2\pi$ from an angle as often as you like to determine what other angles share the same terminal side.

**Example 4:** Find two positive and two negative coterminal angles for $\theta = \frac{3\pi}{4}$.

Add $2\pi$ to $\theta$ twice; each result will be a positive coterminal angle.

$$\frac{3\pi}{4} + 2\pi = \frac{11\pi}{4}$$
$$\frac{11\pi}{4} + 2\pi = \frac{19\pi}{4}$$

Subtract $2\pi$ from $\theta$ twice; each result will be a negative coterminal angle.

$$\frac{3\pi}{4} - 2\pi = -\frac{5\pi}{4}$$
$$-\frac{5\pi}{4} - 2\pi = -\frac{13\pi}{4}$$

## The Unit Circle

The **unit circle** is just a circle, centered at the origin, which has radius 1. By examining the intersection point of the unit circle and the terminal side of an angle in standard position, you can easily determine the values of *cosine* and *sine* for that angle.

As you learned in geometry, *cosine* and *sine* are two of the six trigonometric functions (you'll read about the others in the next section) which are defined as ratios of the lengths of the sides of right triangles. Specifically, the **cosine** of an angle (abbreviated "cos") is defined as the ratio of the adjacent leg to the hypotenuse, and the **sine** (abbreviated "sin") is the ratio of the leg opposite the angle to the hypotenuse. (Remember, the hypotenuse is not considered a leg of the triangle.)

$$\cos\theta = \frac{\text{leg adjacent to } \theta}{\text{hypotenuse}}$$

$$\sin\theta = \frac{\text{leg opposite } \theta}{\text{hypotenuse}}$$

To calculate cos θ and sin θ, follow these steps:

1. Graph θ in standard position and mark its intersection point *P* with the unit circle.
2. Draw a right triangle connecting *P*, the point on the *x*-axis directly below *P*, and the origin, as pictured in Figure 5-4.
3. Use geometric rules and properties to calculate the lengths of the horizontal and vertical sides of the triangle. This gives you the coordinate value of *P*.
4. Plug the lengths of the legs into the appropriate trigonometric ratios.

**Figure 5-4** The coordinates of *P* provide the values of cosine and sine for that angle.

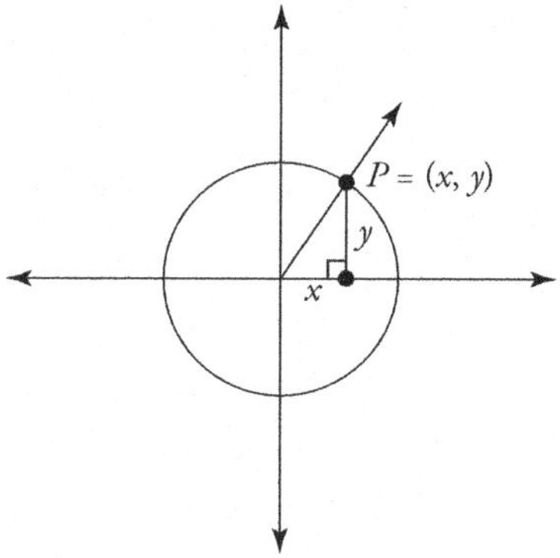


It is not always easy to find the lengths of the sides of the right triangle; in fact, the unit circle is only really useful for quadrantals or radian angles with denominators of 2, 3, 4, and 6. The good news is that you're not expected to generate values of sine and cosine using the unit circle once you understand how it works.

**Example 5:** Determine the unit circle values for $\theta = \frac{\pi}{3}$ by evaluating $\cos\frac{\pi}{3}$ and $\sin\frac{\pi}{3}$.

Draw $\theta = \frac{\pi}{3}$ in standard position so that its terminal side intersects the unit circle at point *P*. Draw a triangle connecting *P*, the point on the *x*-axis directly below *P*, and the origin.

The result is a right triangle containing an angle equivalent to 60°, so you can surmise that it is a 30-60-90 triangle. The legs of such triangles are in fixed proportion to one another, according to geometry. The smallest side must be half the length of the hypotenuse $\left(\frac{1}{2}\cdot 1 = \frac{1}{2}\right)$ and the other leg must be $\sqrt{3}$ times as long as the smallest side, as demonstrated in Figure 5-5.

**Figure 5-5** Use geometric theorems to find the lengths of the triangle's legs.

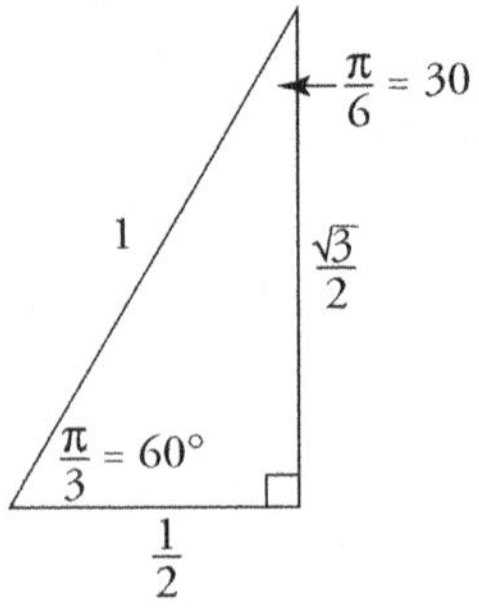


To calculate cos θ and sin θ, plug the lengths of the legs into the trigonometric ratios.

$$\cos\frac{\pi}{3} = \frac{\frac{1}{2}}{1} = \frac{1}{2}$$

$$\sin\frac{\pi}{3} = \frac{\frac{\sqrt{3}}{2}}{2} = \frac{\sqrt{3}}{2}$$

This is a useful exercise because it shows you exactly where the values of cosine and sine originate. However, it's not useful to reinvent the wheel and repeat this process for all common radian angles. In Figure 5-6, all of the tedious calculations have been performed for you. The intersection points between every common terminal side and the unit circle are listed. To determine the cosine of any angle, just take the *x*-value of that coordinate; the sine is the *y*-value of the intersection point.

**Figure 5-6** The complete unit circle.

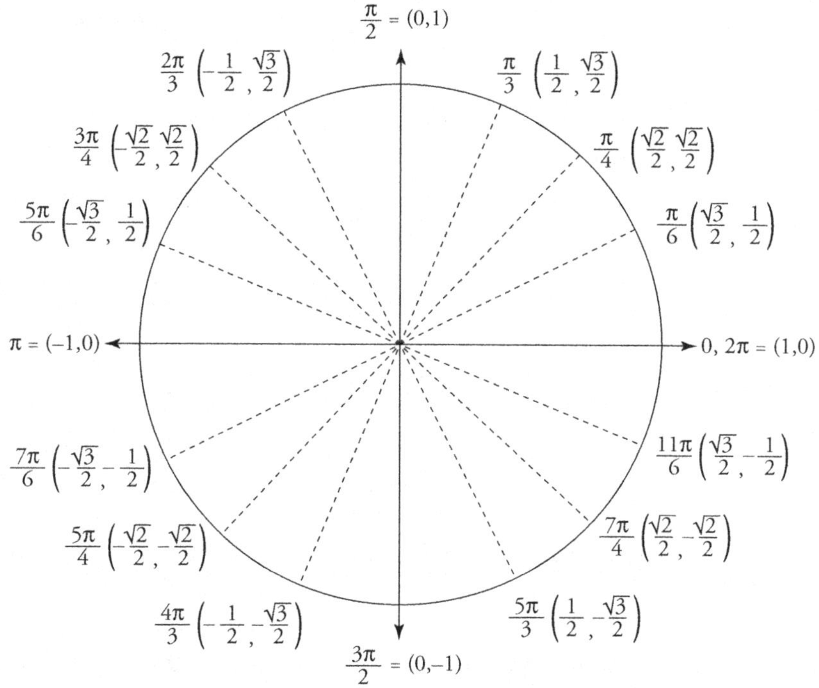


Most instructors require you to memorize this diagram, so you can instantly recite the cosine or sine value of each angle.

## Right Triangle Trigonometry

Just as cosine and sine are defined as the ratios of the sides of a right triangle, so too are the other four major trigonometric functions. In Table 5-1, you'll find all six trigonometric functions, their abbreviations, the ratios which define them, and additional expressions (if available) which are equivalent to the ratio definition of each.

**Table 5-1 The Six Major Trigonometric Functions**

| *Function* | *Abbreviation* | *Right Triangle Ratio* | *Alternate Expressions* |
|---|---|---|---|
| cosine | $\cos\theta$ | $\frac{\text{adjacent leg}}{\text{hypotenuse}}$ | $\frac{1}{\sec\theta}$ |
| sine | $\sin\theta$ | $\frac{\text{opposite leg}}{\text{hypotenuse}}$ | $\frac{1}{\csc\theta}$ |
| tangent | $\tan\theta$ | $\frac{\text{opposite leg}}{\text{adjacent leg}}$ | $\frac{\sin\theta}{\cos\theta}, \frac{1}{\cot\theta}$ |
| cotangent | $\cot\theta$ | $\frac{\text{adjacent leg}}{\text{opposite leg}}$ | $\frac{\cos\theta}{\sin\theta}, \frac{1}{\tan\theta}$ |
| secant | $\sec\theta$ | $\frac{\text{hypotenuse}}{\text{adjacent leg}}$ | $\frac{1}{\cos\theta}$ |
| cosecant | $\csc\theta$ | $\frac{\text{hypotenuse}}{\text{opposite leg}}$ | $\frac{1}{\sin\theta}$ |

There are a few things you should notice about the six trigonometric functions.

- Tangent is defined both as a ratio and as the quotient of sine and cosine. Therefore, if you're given the values of $\cos\theta$ and $\sin\theta$, you can evaluate $\tan\theta$ easily.
- Cotangent, secant, and cosecant are *reciprocal functions,* since they are defined as the reciprocals of the other three functions.
- *Co*secant is not the reciprocal of *co*sine, just as *s*ecant is not the reciprocal of *s*ine. Because the word pairs start the same, students sometimes assume they are reciprocal functions.

Here are the steps to calculate the values of these six trigonometric functions for acute angles within right triangles.

1. Draw a diagram based on the given information.
2. Use the Pythagorean Theorem to find the value of any side whose length is not given.
3. Determine the sine and cosine of the given angle.
4. Evaluate any or all of the other four trigonometric functions based on the cosine and sine values.

**Example 6:** If $\cos\theta = \frac{3}{5}$, and $\theta$ is an acute angle, evaluate the other five trigonometric functions for $\theta$.

Since cosine represents the ratio of the adjacent leg to the hypotenuse, you know the leg next to $\theta$ has length 3, and the hypotenuse has length 5, as shown in Figure 5-7.

**Figure 5-7** The given value of cosine allows you to create a diagram of a right triangle containing $\theta$.

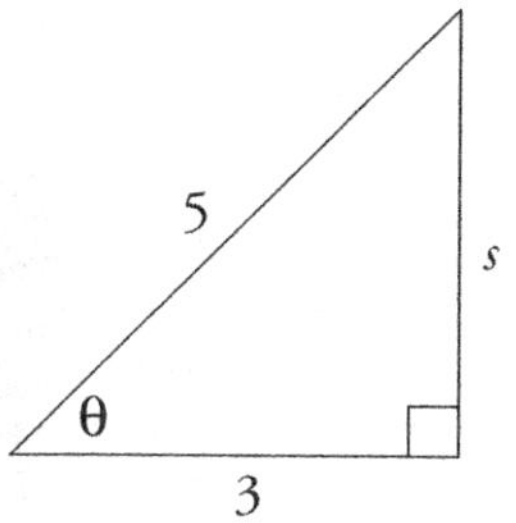


Find $s$ using the Pythagorean Theorem.

$$s^2 = 5^2 - 3^2 = 25 - 9 = 16$$

$$s = 4$$

Now that you know the leg opposite $\theta$ has length 4, $\sin\theta$ must equal $\frac{4}{5}$. Since tangent is the ratio of the opposite to the adjacent leg, $\tan\theta = \frac{4}{3}$. By the way, you can also find tangent using its alternate definition:

$$\tan\theta = \frac{\sin\theta}{\cos\theta} = \frac{\frac{4}{5}}{\frac{3}{5}} = \frac{4}{5} \cdot \frac{5}{3} = \frac{4}{3}$$

The final three trigonometric functions (secant, cosecant, and cotangent) are just reciprocals of the cosine, sine, and tangent, respectively.

$$\sec\theta = \frac{5}{3} \quad \csc\theta = \frac{5}{4} \quad \cot\theta = \frac{3}{4}$$

If you're given an angle and the length of one side of a right triangle, you can use trigonometric functions to find the lengths of other sides in the triangle. In these cases, you'll need to use either a calculator or a table containing trigonometric values in order to complete the problem.

**Example 7:** You are lying on the ground so that your eyes are 25 feet away from the base of a tree. When you look at the top of the tree, the angle of elevation of your vision is 32°. How tall is the tree?

Assuming the tree grows perpendicular to the ground, a right triangle is formed. Draw a diagram like Figure 5-8 that includes all of the information you're given. By the way, an angle of elevation is usually formed by a line of sight and a horizontal reference plane (usually the ground).

**Figure 5-8** The diagram describing Example 7.

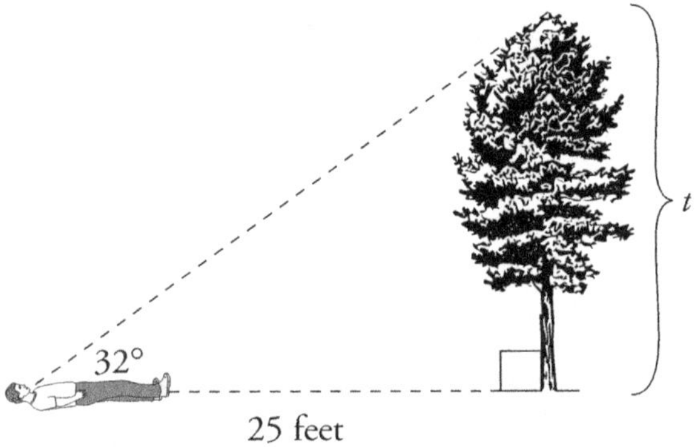


You're given nothing about the hypotenuse, so pick a trigonometric ratio that includes the adjacent leg (which you're given) and the opposite leg (which you're trying to find); tangent will work.

$$\tan 32° = \frac{\text{opposite leg}}{\text{adjacent leg}}$$

$$\tan 32° = \frac{t}{25}$$

Evaluate tan 32° using a calculator and solve for $t$.

$$t = 25 \cdot \tan 32° \approx 15.622 \text{ feet}$$

Make sure your calculator is set for "degrees" mode, not "radians," to match the angle measurement method used in the problem.

## Oblique Triangle Trigonometry

You can also calculate trigonometric function values for triangles which do not contain a right angle, which are called **oblique triangles**. However, the process involves slightly more detective work.

## Reference angles

If you are trying to evaluate trigonometric functions for an angle that's not acute, you can't apply the appropriate ratios right away. That's because a non-acute angle cannot be part of a right triangle, and all of the trigonometric functions are defined as ratios of the sides of right triangles only.

Therefore, you'll need to create an acute angle, called a **reference angle**, to help you evaluate trigonometric function values of large angles. Although reference angles don't have the same measure as the original angles, they do have the same trigonometric values. There are four basic types of *reference angles*, one for each of the four quadrants of the coordinate system, as you can see in Figure 5-9. Choose the one in the same quadrant as the terminal side of the non-acute angle.

**Figure 5-9** Each quadrant has its own reference angle, each of which forms a right triangle with the *x*-axis.

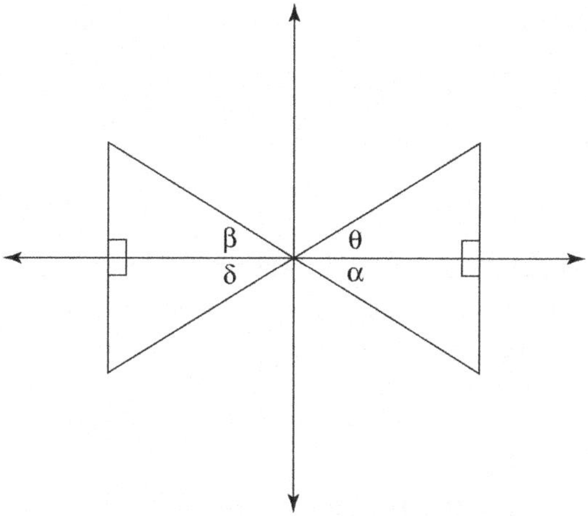


**Example 8:** Calculate the reference angle for $\gamma = \frac{7\pi}{6}$.

If drawn in standard position, the terminal side of $\frac{7\pi}{6}$ falls in the third (lower left-hand) quadrant. The right triangle for all third quadrant angles, according to Figure 5-9, will contain reference angle δ, so your job is to calculate δ.

As you can see in Figure 5-10, while γ extends from the first quadrant all the way into the third quadrant, δ equals just the portion of γ in the third quadrant.

**Figure 5-10** Angle $\gamma$ and its reference angle $\delta$ from Example 8(a).

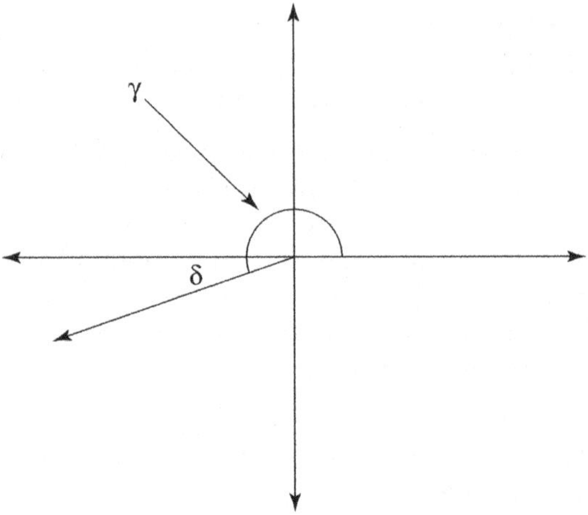


To find $\delta$, subtract the radians contained in the first and second quadrant:

$$\delta = \gamma - \pi$$
$$\delta = \frac{7\pi}{6} - \frac{6\pi}{6}$$
$$\delta = \frac{\pi}{6}$$

## Calculating trigonometric ratios

Use reference angles and the right triangles which contain them to calculate trigonometric ratios for non-acute angles. Occasionally, you'll be given information about the sign of an angle's trigonometric ratio to help you determine in which quadrant its terminal side is located. In such cases, refer to Table 5-2.

**Table 5-2 Signs of the Trigonometric Functions in the Four Quadrants**

| *Function* | *I* | *II* | *III* | *IV* |
|---|---|---|---|---|
| cosine | + | – | – | + |
| sine | + | + | – | – |
| tangent | + | – | + | – |
| secant | + | – | – | + |
| cosecant | + | + | – | – |
| cotangent | + | – | + | – |

**Example 9:** If $\cos \gamma = -\frac{2}{3}$ and $\tan \gamma < 0$, evaluate $\csc \gamma$.

First, determine the type of reference angle needed; in which quadrant does the terminal side of $\gamma$ fall? You know that cosine is negative there, so according to Table 5-2, it must be either quadrant II or III. Note that the problem also tells you that tangent is negative there, so it must be quadrant II.

Draw a second quadrant reference angle and its accompanying right triangle. You know the adjacent leg has length 2 and the hypotenuse has length 3 (since you're given the cosine function), but where does the negative sign go on the triangle? Note that the hypotenuse should never be labeled negative, only the horizontal or vertical sides of the right triangle. Since you're in the second quadrant, $x$ is negative and $y$ is positive, so therefore, the adjacent side should be labeled $-2$, as shown in Figure 5-11. The final side's length, $\sqrt{5}$, can be found via the Pythagorean Theorem.

**Figure 5-11** The reference angle, β, for Example 9's angle γ and its accompanying right triangle. The figure is not to scale, but since you don't know the scale when you first sketch it, that's to be expected.

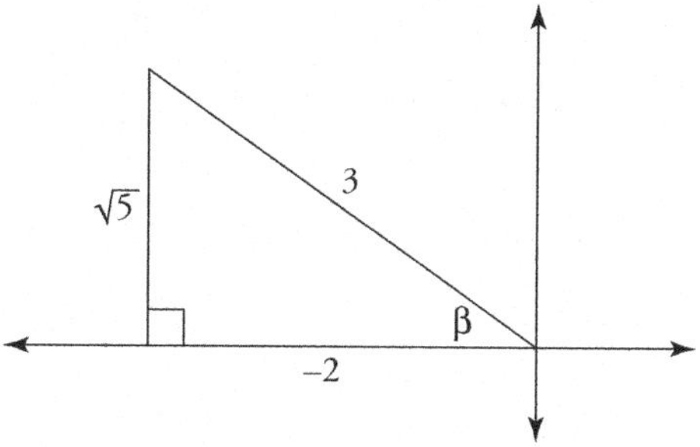


Even though $\gamma$ was not an acute angle, you have now designed a right triangle with acute reference angle $\beta$, which will produce the same trigonometric values. Therefore, $\csc \gamma = \csc \beta$.

Remember that cosecant is the reciprocal of sine.

$$\sin\beta = \frac{\sqrt{5}}{3}$$

$$\csc\beta = \frac{3}{\sqrt{5}}$$

# Graphs of Sine and Cosine

In Chapter 2, you were introduced to eight basic function graphs which, when shifted, stretched, and reflected, constituted the vast majority of graphs in precalculus. Now that you possess a basic understanding of trigonometric principles, you need to add the graphs of the trigonometric functions to your repertoire. Although you'll be expected to graph each function and identify the key characteristics of each, you will work most with the graphs of sine and cosine.

## Periodic graphs

All of the trigonometric functions' graphs are **periodic**, which means that the graph will repeat itself infinitely (both as $x$ gets infinitely positive and infinitely negative) after some fixed interval **period** on the $x$-axis. The *period* of sine is $2\pi$. This makes sense, if you consider that both sine and cosine get their value from the unit circle.

The graph of sine is sometimes referred to as a *sine wave*, because of its shape. Figure 5-12 shows one period of sine, from $\theta = 0$ to $\theta = 2\pi$. Each of the points indicated on the graph represents a value from the unit circle.

**Figure 5-12** One period of the graph $y = \sin x$.

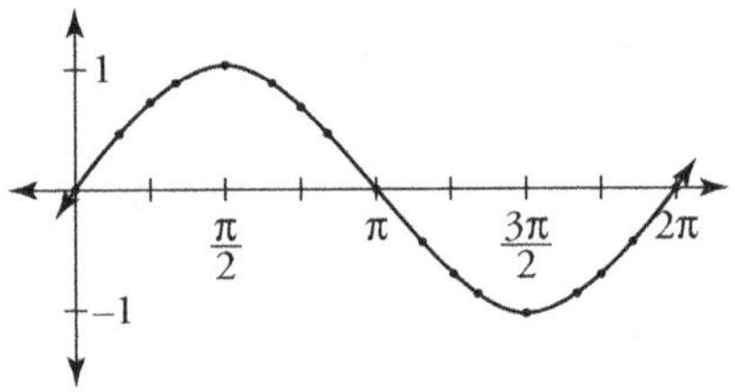


The graph of cosine is identical to the graph of sine shifted $\frac{\pi}{2}$ units to the right, as you can see in Figure 5-13. Since its values are also based on the unit circle, its period is also $2\pi$.

**Figure 5-13** One period of the graph $y = \cos x$.

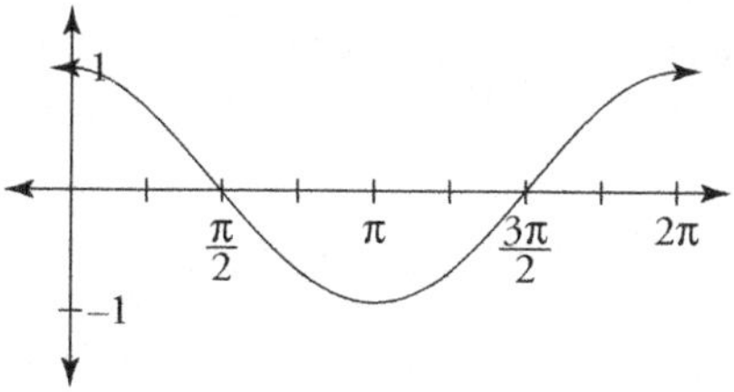


Because both graphs are periodic, you can evaluate any angle coterminal to an angle on the unit circle without using a calculator.

**Example 9:** Evaluate the trigonometric functions.

(a) $\cos \frac{15\pi}{2}$

Find an angle on the interval $[0,2\pi)$ which is coterminal with $\frac{15\pi}{2}$ by continuously subtracting $2\pi$ until you find one. The coterminal angle on the unit circle will be $\frac{3\pi}{2}$, and you already should know its cosine value. Therefore,

$$\cos \frac{15\pi}{2} = \cos \frac{3\pi}{2} = 0$$

(b) $\sin -\frac{5\pi}{6}$

This time *add* $2\pi$ to get a coterminal angle on the interval $[0,2\pi)$; its sine value will be the same.

$$-\frac{5\pi}{6} + 2\pi = \frac{7\pi}{6}$$

$$\sin \frac{7\pi}{6} = -\frac{1}{2}$$

### Transforming sine and cosine

Most of the shifts, reflections, and stretching you did to functions in Chapter 2 still apply to the graphs of trigonometric functions. However, since these functions are periodic, they require a few special considerations. The functions

$$f(x) = a\cos(bx + c) + d \text{ and } g(x) = a\sin(bx + c) + d$$

will be altered from the original graphs of $f(x) = \cos x$ and $g(x) = \sin x$ by the coefficients as follows:

- *a*: The graph, instead of reaching a maximum height of 1 and a minimum height of –1, will now reach maximum and minimum heights of *a* and –*a*, respectively. In other words, the graph is stretched vertically by a factor of *a*. (If $a < 0$, the function will also be reflected across the *x*-axis.)

  The value $|a|$ is called the **amplitude** of the function. If given only the graph of the function, you can determine its amplitude by subtracting the function's minimum value from its maximum value and dividing by 2.

- *b*: This is the number of times the graph will repeat itself in what used to be its original period, $2\pi$. You can also use *b* to find the period of the transformed function.

$$\text{new period} = \frac{\text{old period}}{b}$$

- *c*: Just like the function transformations in Chapter 2, a *c* value causes a horizontal function shift. If $c > 0$, the shift is to the left, and if $c < 0$, the shift is to the right.

- *d*: If $d > 0$, the graph will be shifted *d* units up, and if $d < 0$, the graph is shifted *d* units down.

If there is more than one transforming coefficient present, apply the transformations in this order: *b*, *a*, *c*, *d*.

**Example 10:** Sketch the following graphs.

(a) $f(x) = -3\sin(2x) + 1$

Begin with $b = 2$; it tells you that exactly two full periods of the graph will fit into $2\pi$, the period for $y = \sin x$, because the period of $f$ equals $\frac{2\pi}{2} = \pi$. (This squashes the function horizontally toward the origin.) The coefficient $a = -3$ will reflect the sine graph across the *x*-axis and stretch it upwards and downwards to a height of 3 and –3, respectively. Finally, $d = 1$ moves the entire graph up one unit. The original graph of $y = \sin x$ and the transformed version, $f(x)$, are shown in Figure 5-14.

(b) $g(x) = \frac{1}{4}\cos(x - \pi)$

To get the graph of *g*, start with the graph of $y = \cos x$, give it an amplitude of $\frac{1}{4}$ (which squishes its height) and move the entire graph $\pi$ units to the right. See Figure 5-15.

**Figure 5-14** The graph of $f(x)$, the answer to Example 10(a), and the pre-transformed graph of $y = \sin x$.

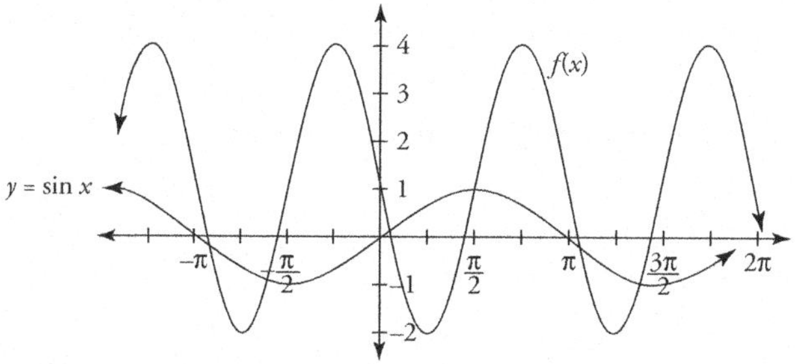


**Figure 5-15** The graph of $g(x)$, the answer to Example 10(b), and the pre-transformed graph of $y = \cos x$.

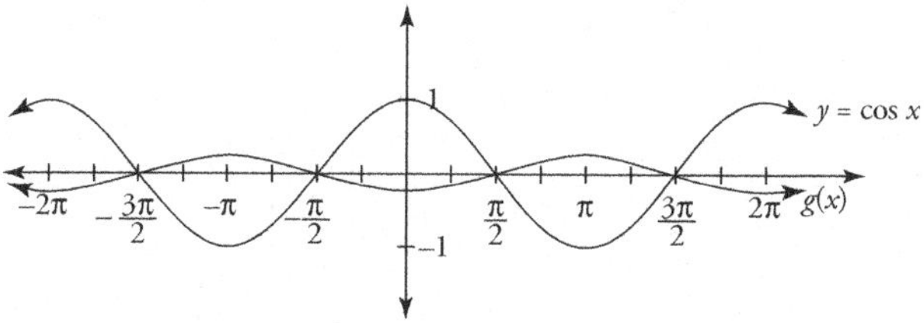


## Other Trigonometric Function Graphs

Because the remaining trigonometric functions (tangent, secant, cosecant, and cotangent) are all defined as fractions, their graphs contain asymptotes (at the values which make their denominators 0). You may be asked to graph a transformation of any of these functions, and the process undertaken is identical to the process of graphing transformations of sine and cosine.

Here are each of the remaining trigonometric graphs, and their important characteristics.

- **$y = \tan \theta$:** Since tangent is defined as $\frac{\sin \theta}{\cos \theta}$, the graph of tangent has an asymptote wherever the graph of $\cos \theta$ has an $x$-intercept. In other words, the graph has asymptotes $y = \frac{k\pi}{2}$, where $k$ is an odd integer. The roots of tangent occur at all $x = n\pi$, where $n$ is an integer. Note that the period of tangent is $\pi$, not $2\pi$ like sine and cosine.

- $y = \cot \theta$**:** Like tangent, the period of cotangent is $\pi$. However, the roots for cotangent occur where tangent has asymptotes, and vice versa. Furthermore, the graph of cotangent looks like the graph of tangent reflected about the $y$-axis. See Figure 5-16.

**Figure 5-16** The graphs of $y = \tan \theta$ and $y = \cot \theta$.

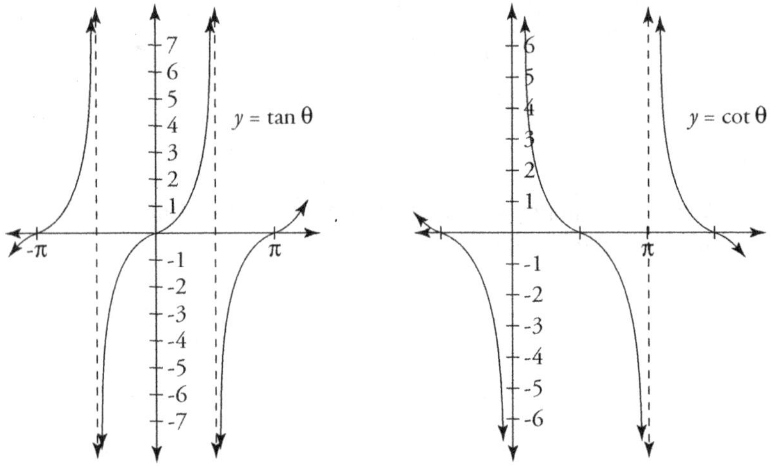


- $y = \sec \theta$**:** Since secant is defined as $\frac{1}{\cos \theta}$, it has the same denominator as tangent. Therefore, they share the same asymptotes as well; secant, however, has no roots. Its graph blossoms from the highest and lowest points of its reciprocal function, cosine. For example, since $y = \cos x$ has relative maximum point $(0,1)$ and minimum point $(\pi,-1)$, those are the points from which the graph of secant springs.
- $y = \csc \theta$**:** Cosecant shares the same asymptotes as cotangent. Like its sister function secant, it has no roots, and its graph springs from the relative maximum and minimum points of its reciprocal function, $y = \sin x$. See Figure 5-17.

**Figure 5-17** The graphs of $y = \sec\theta$ and $y = \csc\theta$.

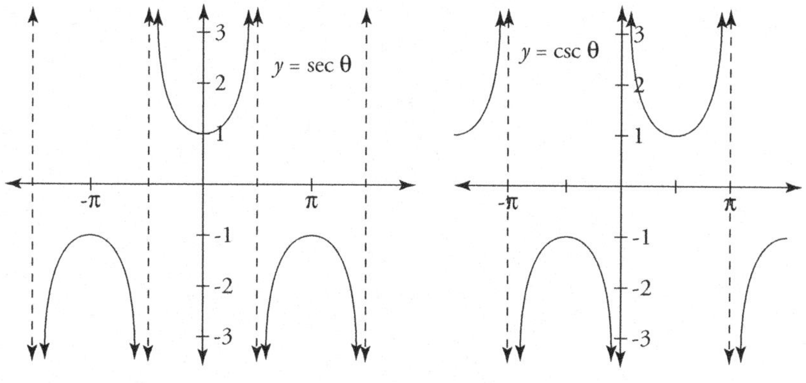


# Inverse Trigonometric Functions

Since all functions must pass the horizontal line test if they are to possess inverses, it would seem that trigonometric functions (since they are periodic and are therefore by definition going to repeat the same function heights over and over) are not one-to-one and cannot have inverse functions.

While it's true that $y = \sin x$ is not one-to-one, the portion of the graph on the $x$-interval $\left[-\frac{\pi}{2}, \frac{\pi}{2}\right]$ is. (You can visually verify that the relatively small piece of the sine graph does pass the horizontal line test in Figure 5-18.) Therefore, you can construct an inverse function for that **restricted sine function**. The inverse function, also pictured in Figure 5-18, is denoted "$\sin^{-1} x$" (read "sine inverse of $x$") or "arcsin $x$" (read "arc sine of $x$").

You are more likely to see the "arcsin" notation than the "$\sin^{-1}$" notation, because the latter expression is sometimes erroneously interpreted as "the reciprocal of sine."

$$\sin^{-1} x \neq \frac{1}{\sin x}$$

Essentially, the expression "arcsin $c$" asks the question "What angle $\theta$ on the interval $\left[-\frac{\pi}{2}, \frac{\pi}{2}\right]$ has a sine value of $c$?" Unlike the sine function, which uses angles as inputs, the arcsin function will always return angles as outputs.

**Figure 5-18** The restricted sine function and its inverse, arcsin $x$. Note that the domain of sin $x$ and the range of arcsin $x$ are both restricted to $\left[-\frac{\pi}{2}, \frac{\pi}{2}\right]$.

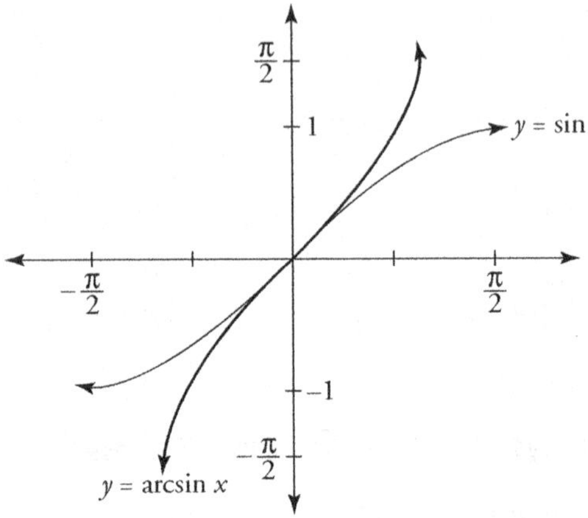


The other five trigonometric functions can also be restricted so that they, too, have inverses whose ranges are in turn restricted; arccsc θ and arctan θ share the same range restriction as arcsin: $-\frac{\pi}{2} \le \theta \le \frac{\pi}{2}$. The range restriction for arccos θ, arcsec θ, and arccot θ has the same interval length but has different endpoints: $0 \le \theta \le \pi$.

**Example 11:** Evaluate the following expressions.

(a) $\arcsin\left(\sin\frac{3\pi}{2}\right)$

Evaluate the inner function first.

$$\arcsin(-1)$$

Where is the sine function equal to –1? At $\theta = \frac{3\pi}{2}$, but that does not fall within the restricted range. Therefore, find a coterminal angle which does fall in that range by subtracting $2\pi$ from $\frac{3\pi}{2}$.

$$\frac{3\pi}{2} - 2\pi = -\frac{\pi}{2}$$

(b) $\tan\left(\text{arcsec} -\frac{7}{3}\right)$

Since the range of arcsecant is $0 \le \theta \le \pi$, it applies only to angles in quadrant I or II. Secant is only negative in quadrant II, so you can rewrite the expression as tan θ, and draw a right triangle with a second quadrant reference angle to visualize θ, as shown in Figure 5-19.

**Figure 5-19** A diagram of θ, based on the information in Example 11(b) and the Pythagorean Theorem.

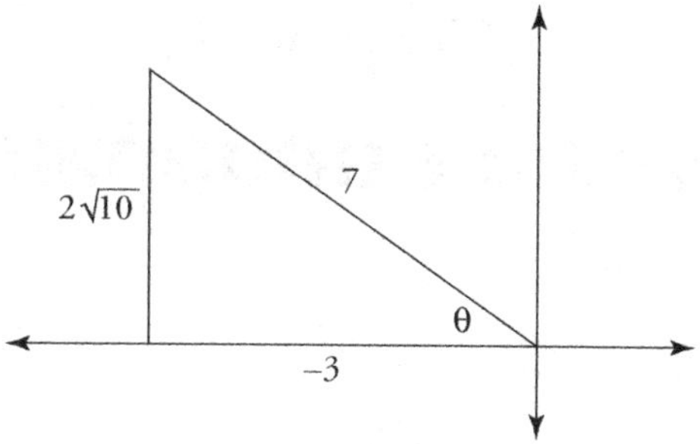


Although you don't know exactly what $\theta$ is, you can still revaluate tan $\theta$ and complete the problem, since tangent is equal to the opposite leg divided by the adjacent one.

$$\tan\left(\text{arcsec}-\frac{7}{3}\right)=\tan\theta=-\frac{2\sqrt{10}}{3}$$

## Chapter Checkout

### Q&A

1. What is the greatest negative coterminal angle for $\theta = \frac{\pi}{6}$?
2. Convert 315° to radians.
3. If $\theta$ is in standard position and its terminal side passes through the point (–1,–5), evaluate sin $\theta$.
4. Give the amplitude and period of $f(x) = -4\cos\left(\frac{\pi}{3}\theta\right) + 5$.
5. Evaluate $\arccos\left(\sin-\frac{\pi}{3}\right)$.

**Answers: 1.** $-\frac{11\pi}{6}$ **2.** $\frac{7\pi}{4}$ **3.** $-\frac{5}{\sqrt{26}}$ **4.** amp = 4; per = $6\pi$ **5.** $\frac{5\pi}{6}$

# Chapter 6
# ANALYTIC TRIGONOMETRY

## Chapter Check-In

- ❑ Defining trigonometric identities
- ❑ Proving statements using identities
- ❑ Solving equations involving trigonometry
- ❑ Applying double- and half-angle identities
- ❑ Understanding the Laws of Sines and Cosines

Whereas Chapter 5 focused exclusively upon introducing the elements of trigonometry, this chapter will help you apply them appropriately in the situations for which trigonometry is most used in precalculus. You'll begin with a brief foray in manipulating trigonometric expressions, then solve trigonometric equations, and eventually apply mathematical laws to evaluate angles and sides of oblique triangles given very little information.

This is the culmination of trigonometry for your purposes, because it shows that the concepts you've learned thus far are not only useful for situations involving right triangles or angles in standard position, but that information about all triangles can be obtained by applying trigonometric functions.

## Trigonometric Identities

Although **trigonometric identities** look a lot like trigonometric equations, they differ in one key characteristic. Whereas an equation may have one, two, or a handful of solutions, an *identity* is true no matter what value is plugged in for the variable.

For example, the equation $x + 5 = 9$ is an equation with one solution ($x = 4$). However, the equation $2(x + 1) = 2x + 2$ falls under the category of *identity* because it is true for any real number $x$. Granted, it's not a very interesting or useful identity, because it cannot be applied in very many situations.

## Four types of identities

There are four main categories of trigonometric identities with which you should become familiar. In fact, you should have each identity memorized, so that you can apply them without referring to this list.

- **Reciprocal identities:** Remember that the cotangent, cosecant, and secant functions are all defined as the reciprocals of the other three trigonometric functions.

$$\cot\theta = \frac{1}{\tan\theta} = \frac{\cos\theta}{\sin\theta}$$
$$\sec\theta = \frac{1}{\cos\theta}$$
$$\csc\theta = \frac{1}{\sin\theta}$$

The reciprocal relationship is a two-way street, however, so you can take the reciprocal of both sides of each of the above equations, and the result is another list of identities.

$$\tan\theta = \frac{1}{\cot\theta} = \frac{\sin\theta}{\cos\theta}$$
$$\cos\theta = \frac{1}{\sec\theta}$$
$$\sin\theta = \frac{1}{\csc\theta}$$

- **Pythagorean identities:** Since the cosine and sine of an angle on the unit circle represent the lengths of the legs of a right triangle, you know that the following is true for all angles:

$$\cos^2\theta + \sin^2\theta = 1$$

Be aware that $\cos^2\theta$ means the exact same thing as $(\cos\theta)^2$ and $\sin^2\theta = (\sin\theta)^2$; it is simply shorthand notation that doesn't require as many parentheses. If you divide that identity by either $\cos^2\theta$ or $\sin^2\theta$, you get one of the other two Pythagorean identities.

$$1 + \tan^2\theta = \sec^2\theta$$
$$1 + \cot^2\theta = \csc^2\theta$$

- **Sign identities:** Based on the symmetry exhibited in their graphs, you can tell that sine, cosecant, tangent, and cotangent are odd functions. Since an odd function $f(x)$ has the property $f(-x) = -f(x)$, the following statements are true:

$$\sin(-\theta) = -\sin\theta$$
$$\csc(-\theta) = -\csc\theta$$
$$\tan(-\theta) = -\tan\theta$$
$$\cot(-\theta) = -\cot\theta$$

The only even trigonometric functions are cosine and secant, which exhibit this property:

$$\cos(-\theta) = \cos\theta$$
$$\sec(-\theta) = \sec\theta$$

- **Cofunction identities: Cofunctions** are trigonometric function pairs which differ only in the presence or absence of the prefix "co". For example, tangent and *co*tangent are *cofunctions*, as are sine and *co*sine.

If $f(\theta)$ and $g(\theta)$ are *cofunctions*, then

$$f\left(\frac{\pi}{2} - \theta\right) = g(\theta)$$
$$g\left(\frac{\pi}{2} - \theta\right) = f(\theta)$$

For example, since sine and cosine are cofunctions, then the following two statements are valid identities:

$$\sin\left(\frac{\pi}{2} - \theta\right) = \cos\theta$$
$$\cos\left(\frac{\pi}{2} - \theta\right) = \sin\theta$$

## Simplifying expressions with identities

Trigonometric identities, when combined with basic algebraic skills such as factoring, can be used to rewrite expressions in simplified form.

**Example 1:** Simplify each of the following expressions.

(a) $\tan(-t) \cdot \cos(-t)$

Rewrite using the sign identities.

$$-\tan t \cdot \cos t$$

Express tangent as a quotient.

$$-\frac{\sin t}{\cos t} \cdot \cos t$$

Cancel out the $\cos t$ terms.

$$-\sin t$$

(b) $\csc^2 \theta - \csc^2 \theta \cos^2 \theta$

Factor $\csc^2 \theta$ out of both terms.

$$\csc^2 \theta(1 - \cos^2 \theta)$$

Consider the Pythagorean identity containing cosine and sine. If you were to subtract $\cos^2 \theta$ from both sides of that identity, it would still hold true, and it would result in

$$\cos^2 \theta + \sin^2 \theta = 1$$
$$\sin^2 \theta = 1 - \cos^2 \theta$$

The quantity on the right side of the equation appears in the factored form of this problem, so substitute $\sin^2 \theta$ for it.

$$\csc^2 \theta(1 - \cos^2 \theta)$$
$$= \csc^2 \theta(\sin^2 \theta)$$

Rewrite using a reciprocal identity.

$$= \frac{1}{\sin^2 \theta} \cdot \sin^2 \theta$$
$$= 1$$

(c) $\dfrac{\cot\left(\frac{\pi}{2} - x\right)}{\sec^2 x - 1}$

Rewrite the numerator using a cofunction identity. You can subtract 1 from both sides of the Pythagorean identity containing tangent and secant to substitute in for the denominator, similar to the substitution you made in part (b).

$$\frac{\tan x}{\tan^2 x}$$

Simplify the fraction and apply a reciprocal identity to finish.

$$\frac{1}{\tan x} = \cot x$$

## Proving Trigonometric Identities

The skills required to simplify trigonometric expressions can be further utilized to prove that equations are actually trigonometric identities. Students take different approaches. Some manipulate both sides of the given equation until they can tell that it is obviously true. Others work with only one side of the equation and try to get it to match the other. Whichever your strategy, you know you've proven the identity when you reach one of the following:

- A known trigonometric identity (such as a Pythagorean identity)
- An obviously true statement, such as "$\sin\theta = \sin\theta$"

**Example 2:** Prove the following trigonometric identities.

(a) $\frac{\sin x}{1-\cos x} = \frac{1+\cos x}{\sin x}$

Cross-multiply to eliminate the fractions.

$$\sin^2 x = 1 - \cos^2 x$$

Add $\cos^2 x$ to both sides to reach a known identity, and therefore accomplish your goal.

$$\cos^2 x + \sin^2 x = 1$$

(b) $\csc\theta\cot\theta\sec\theta = \sec^2\left(\frac{\pi}{2} - \theta\right)$

Rewrite the right side using a cofunction identity.

$$\csc\theta\cot\theta\sec\theta = \csc^2\theta$$

Rewrite the left side using reciprocal identities.

$$\frac{1}{\sin\theta}\cdot\frac{\cos\theta}{\sin\theta}\cdot\frac{1}{\cos\theta} = \csc^2\theta$$

The $\cos\theta$ terms will cancel out, leaving a statement that is true according to a reciprocal identity.

$$\frac{1}{\sin^2\theta} = \csc^2\theta$$

(c) $1 - \frac{\sin\left(\frac{\pi}{2} - x\right)}{\sec(-x)} = \sin^2 x$

Leave the right side alone and simplify the left side; begin with the fraction. Rewrite its numerator with a cofunction identity and its denominator with a sign identity.

$$1 - \frac{\cos x}{\sec x} = \sin^2 x$$

Rewrite secant using a reciprocal identity.

$$1 - \frac{\cos x}{\frac{1}{\cos x}} = \sin^2 x$$

Simplify the complex fraction.

$$1 - \cos^2 x = \sin^2 x$$

$$\cos^2 + \sin^2 x = 1$$

(d) $\sec^4 \theta - \tan^4 \theta = \dfrac{1 + \sin^2 \theta}{\cos^2 \theta}$

The left side is the difference of perfect squares; factor it.

$$(\sec^2 \theta + \tan^2 \theta)(\sec^2 \theta - \tan^2 \theta) = \frac{1 + \sin^2 \theta}{\cos^2 \theta}$$

According to a Pythagorean identity, $\sec^2 \theta - \tan^2 \theta = 1$, so substitute that value.

$$(\sec^2 \theta + \tan^2 \theta)(1) = \frac{1 + \sin^2 \theta}{\cos^2 \theta}$$

Rewrite the right side of the equation as the sum of two fractions.

$$\sec^2 \theta + \tan^2 \theta = \frac{1}{\cos^2 \theta} + \frac{\sin^2 \theta}{\cos^2 \theta}$$

Rewrite the left side of the equation using reciprocal identities.

$$\frac{1}{\cos^2 \theta} + \frac{\sin^2 \theta}{\cos^2 \theta} = \frac{1}{\cos^2 \theta} + \frac{\sin^2 \theta}{\cos^2 \theta}$$

## Solving Trigonometric Equations

Trigonometric equations don't require many special techniques to solve. In fact, you'll use the same methods you did with ordinary, polynomial equations, with one important exception: You have to write your solution in the very specific way called for by the problem. Although an equation will always have the same solutions, there are different ways of writing those solutions based on how the problem is worded.

Consider this basic trigonometric equation:

$$\sin \theta = 0$$

Your answer for this, as with all trigonometric equations, will be an angle or angles that make the statement true. Here are the different solutions you should give to that equation depending upon the problem's instructions.

- **Exact solution:** If you are asked to provide the *exact solution*, you should pay special attention to the restricted domain of the inverse trigonometric function. To solve the equation $\sin \theta = 0$, you should take the arcsin of both sides.

  $$\arcsin (\sin \theta) = \arcsin 0$$

  The left side will simplify to $\theta$; when you evaluate the right side, remember you can only give *one* output, since arcsine is a function, and that solution must fall within the interval $-\frac{\pi}{2} \leq \theta \leq \frac{\pi}{2}$, as discussed in Chapter 5.

  $$\theta = 0$$

- **Specified solution:** Sometimes, the problem will ask you to provide solutions on a predefined interval, usually $[0,2\pi)$. If that interval is specified in this problem, you should give two solutions: $\theta = 0, \pi$, since they both make the statement true. You're not breaking the rules of inverse functions, you're just choosing to ignore them. Even though arcsin 0 should only return one value, there's no denying that both $\theta = 0$ and $\theta = \pi$ make the equation true.

- **General solution:** A general solution includes an infinite number of angles, because it considers all the angles coterminal to the solution as valid answers, as well. You already know that both 0 and $\pi$ are valid solutions to the equation $\sin \theta = 0$. To make that into a general solution, include all possible coterminal angles of each individual solution by adding multiples of the trigonometric function's period like this:

  $$\theta = 0 + 2\pi(n)$$

  $$\theta = \pi + 2\pi(n)$$

  where $n$ is an integer. In fact, you can write this solution even simpler, since every viable answer is exactly $\pi$ more than another.

  $$\theta = \pi + n\pi, \text{ if } n \text{ is an integer}$$

In the problems that follow, the directions will request different kinds of solutions, but remember that any type of solution can be requested for any type of trigonometric equation.

## Simple equations

These equations looks just like linear equations, except that they contain trigonometric functions.

**Example 3:** Give all solutions to the equation on the interval $[0,2\pi)$.

$$\sqrt{3}\sec\theta - 2 = 0$$

Isolate sec θ on the left side of the equation.

$$\sec\theta = \frac{2}{\sqrt{3}}$$

Take the reciprocal of both sides of the equation.

$$\cos\theta = \frac{\sqrt{3}}{2}$$

According to the unit circle, the angles which have that cosine are $\theta = \frac{\pi}{6}, \frac{11\pi}{6}$.

## Quadratic equations

Just like polynomials, these can be solved via factoring or, if necessary, the quadratic formula.

**Example 4:** Give the general solution to the equation.

$$\tan^2 x + 4\tan x = 5$$

Move all terms to the left side and factor the equation.

$$\tan^2 x + 4\tan x - 5 = 0$$

$$(\tan x + 5)(\tan x - 1) = 0$$

Set each factor equal to 0 and solve. Since tangent has a period of $\pi$, find the solutions on $[0,\pi]$; you'll create the general solution by adding $n\pi$ to these answers.

$$\tan x = -5 \quad \text{or} \quad \tan x = 1$$

$$x = \arctan(-5) \quad \text{or} \quad x = \frac{\pi}{4}$$

Since both the sine and cosine values of $\frac{\pi}{4}$ are equal (and tangent is defined as the quotient of the two), $\tan \frac{\pi}{4} = 1$. There are no angles on the unit circle whose tangent is –5, so you can either leave your answer as arctan (–5) or rewrite as an approximate decimal (–1.3734). Create the general solution by adding multiples of $\pi$, the period of tangent.

$$x = \arctan(-5) + n\pi$$

$$x = \frac{\pi}{4} + n\pi$$

if $n$ is an integer.

## Equations requiring identities

Before you can solve an equation, you need to make sure it is expressed entirely in terms of one trigonometric function. Example 5 makes use of an identity to reach that goal.

**Example 5:** Give the exact solution to the equation.

$$\sin^2\theta + 3\cos\theta - 3 = 0$$

Rewrite $\sin^2\theta$ as $(1 - \cos^2\theta)$, since they are equal according to a Pythagorean identity.

$$(1 - \cos^2\theta) + 3\cos\theta - 3 = 0$$

Multiply everything by –1 and simplify.

$$-1 + \cos^2\theta - 3\cos\theta + 3 = 0$$

$$\cos^2\theta - 3\cos\theta + 2 = 0$$

Factor the left side, and set each factor equal to 0.

$$(\cos\theta - 2)(\cos\theta - 1) = 0$$

$$\cos\theta = 2 \quad \text{or} \quad \cos\theta = 1$$

The left equation has no solution, because the range of cosine is $-1 \le \cos\theta \le 1$, so it can be discarded. Since you are asked to give the exact solution of $\cos\theta = 1$, make sure to report only answer on the restricted domain of $[0,\pi]$.

$$\theta = 0$$

### Equations requiring squaring

It is not always a straightforward task to write an equation in terms of only one trigonometric function. Pythagorean identities make this easier, but in order to use them, squared trigonometric functions must be present. If they aren't, introduce them by squaring both sides of the equation. Beware that this procedure could introduce false solutions, so you must always check your answers to ensure that they are valid.

**Example 6:** Give the general solution to the equation.

$$\sin x + 1 = \cos x$$

Square both sides of the equation; make sure to use FOIL on the left side.

$$\sin^2 x + 2\sin x + 1 = \cos^2 x$$

Since the only non-squared trigonometric function is sine, eliminate cosine with a Pythagorean identity.

$$\sin^2 x + 2\sin x + 1 = 1 - \sin^2 x$$

$$2\sin^2 x + 2\sin x = 0$$

Factor and solve.

$$2\sin x(\sin x + 1) = 0$$

$$2\sin x = 0 \quad \text{or} \quad \sin x = -1$$

The left equation has solutions 0 and $\pi$, and the right equation has solution $\frac{3\pi}{2}$. However, if you test each, you'll notice that $\pi$ does not work.

$$x = 0 + n(2\pi)$$

$$x = \frac{3\pi}{2} + n(2\pi)$$

if $n$ is an integer.

### Functions of multiple angles

If the trigonometric function contains something other than just a variable such as $x$ or $\theta$, you'll only need to alter the final steps of the problem.

**Example 7:** Give the solutions to the equation on the interval $[0,2\pi)$.

$$2\cos(3\theta) + 1 = 0$$

Isolate $\cos(3\theta)$ on the left side of the equation.

$$\cos(3\theta) = -\frac{1}{2}$$

Usually, you would now list the angles on one period of cosine which have a value of $-\frac{1}{2}$. (They are $\frac{2\pi}{3}$ and $\frac{4\pi}{3}$.) However, since the coefficient of $\theta$ is 3, list *three* times as many angles for a total of 6, lengthening the original list of 2 solutions via the next 2 positive coterminal angles of each.

$$3\theta = \frac{2\pi}{3}, \frac{4\pi}{3}, \frac{8\pi}{3}, \frac{10\pi}{3}, \frac{14\pi}{3}, \frac{16\pi}{3}$$

To finish, divide everything by $\theta$'s coefficient.

$$\theta = \frac{2\pi}{9}, \frac{4\pi}{9}, \frac{8\pi}{9}, \frac{10\pi}{9}, \frac{14\pi}{9}, \frac{16\pi}{9}$$

It was only necessary to write all 6 solutions for this problem because the directions indicated that all solutions on the interval $[0,2\pi)$ be given.

## Sum and Difference Identities

Remember that trigonometric functions are just that: functions. They are not real number values that can be distributed throughout a quantity.

$$\sin(x + y) \neq \sin x + \sin y$$

If you encounter a sum or difference within a trigonometric function (like the left side of the above statement), you'll need to use *sum and difference identities* to transform the expression into an altogether new expression, which will contain only trigonometric functions of single angles. These identities are listed below. Note that each $\pm$ symbol could represent either a + or a –, but that the $\mp$ symbol will then take on the opposite value.

- **Sine:** $\sin(\alpha \pm \beta) = \sin\alpha\cos\beta \pm \cos\alpha\sin\beta$
- **Cosine:** $\cos(\alpha \pm \beta) = \cos\alpha\cos\beta \mp \sin\alpha\sin\beta$
- **Tangent:** $\tan(\alpha \pm \beta) = \dfrac{\tan\alpha \pm \tan\beta}{1 \mp \tan\alpha\tan\beta}$

These identities can be used to calculate trigonometric values for angles not on the unit circle, but they are not limited to this task. In fact, you may see them in trigonometric equations or identity problems. Wherever you see a sum or difference within a trigonometric function, your first task should be to expand that function via sum and difference identities.

**Example 8:** Solve the equation and give all solutions on the interval $[0,2\pi)$.

$$\cos\left(x+\frac{\pi}{4}\right)-\cos\left(x-\frac{\pi}{4}\right)=1$$

Expand both sums within the trigonometric expression. Two of the terms will cancel, so evaluate the trigonometric functions in the remaining terms.

$$\left(\cos x\cos\frac{\pi}{4}-\sin x\sin\frac{\pi}{4}\right)-\left(\cos x\cos\frac{\pi}{4}+\sin x\sin\frac{\pi}{4}\right)=1$$

$$-\sin x\left(\frac{\sqrt{2}}{2}\right)-\sin x\left(\frac{\sqrt{2}}{2}\right)=1$$

$$-\sqrt{2}\sin x=1$$

$$\sin x=-\frac{1}{\sqrt{2}}\cdot\frac{\sqrt{2}}{\sqrt{2}}=-\frac{\sqrt{2}}{2}$$

Sine takes on that function value at $\theta=\frac{5\pi}{4},\frac{7\pi}{4}$.

# Additional Identities

Although the identities discussed thus far this chapter are used most frequently, you will occasionally encounter a few additional types of identities. Like previous identities, these fulfill needs that may arise in very specific (but less common) circumstances.

## Double-angle formulas

Since the majority of the identities you've used so far are written in terms of a single angle ($\sin x$ rather than, for example, $\sin 2x$), converting from double angles to single angles is a useful skill. Double-angle expressions should almost always be rewritten as single-angle expressions immediately if observed in an identity or equation.

- $\sin 2\theta = 2\sin\theta\cos\theta$
- $\cos 2\theta = \cos^2\theta - \sin^2\theta$

  $\cos 2\theta = 2\cos^2\theta - 1$

  $\cos 2\theta = 1 - 2\sin^2\theta$
- $\tan 2\theta = \dfrac{2\tan\theta}{1-\tan^2\theta}$

Note that the cosine double-angle formula contains three different but equivalent expansions, so you have some flexibility when substituting in for it. Decide which to use based on the rest of the problem's content; if it contains both sines and cosines, use the first expansion, but if it contains only one or the other, use the expression that matches.

**Example 9:** Prove the identity.

$$2\cos^5 x \sin x - 2\sin^5 x \cos x = \cos 2x \sin 2x$$

Factor $2\sin x \cos x$ out of the left side of the equation.

$$2\sin x \cos x(\cos^4 x - \sin^4 x) = \cos 2x \sin 2x$$

Factor the difference of perfect squares.

$$2\sin x \cos x(\cos^2 x + \sin^2 x)(\cos^2 x - \sin^2 x) = \cos 2x \sin 2x$$

According to a Pythagorean identity, $\cos^2 x + \sin^2 x = 1$.

$$2\sin x \cos x(1)(\cos^2 x - \sin^2 x) = \cos 2x \sin 2x$$

The left side contains two double-angle identity expansions.

$$(\sin 2x)(\cos 2x) = \cos 2x \sin 2x$$

This is true because of the commutative property of multiplication.

## Half-angle formulas

Just like double-angle formulas, half-angle formulas return expressions in terms of single angles.

- $\cos\frac{\theta}{2} = \pm\sqrt{\frac{1+\cos\theta}{2}}$
- $\sin\frac{\theta}{2} = \pm\sqrt{\frac{1-\cos\theta}{2}}$
- $\tan\frac{\theta}{2} = \pm\frac{\sqrt{1-\cos\theta}}{\sqrt{1+\cos\theta}}$

  $\tan\frac{\theta}{2} = \frac{\sin\theta}{1+\cos\theta}$

  $\tan\frac{\theta}{2} = \frac{1-\cos\theta}{\sin\theta}$

Tangent has three alternate half-angle expansions, two of which will generate the correct sign of your answer. The other three of the five half-angle trigonometric expansions contain a "±" sign, which indicates that you must insert the correct sign, depending upon which quadrant the given angle, $\frac{\theta}{2}$, lies in. For example, if $\frac{\theta}{2}$'s terminal side fell in the second quadrant, you'd know that its cosine and tangent values would be negative and its sine would be positive, just like all second quadrant angles.

**Example 10:** Evaluate $\tan \frac{5\pi}{12}$ using a half-angle identity.

Although $\frac{5\pi}{12}$ is not on the unit circle, $\frac{5\pi}{6}$ is. Therefore, set $\theta = \frac{5\pi}{6}$ in one of the half-angle expansions.

$$\tan\frac{\theta}{2} = \tan\frac{5\pi}{12} = \frac{\sin\left(\frac{5\pi}{6}\right)}{1+\cos\left(\frac{5\pi}{6}\right)}$$

$$= \frac{1/2}{1+\left(-\sqrt{3}/2\right)}$$

$$\approx 3.73205$$

## Sum-product formulas

If an equation or identity can be made simpler by transforming the sum or difference of two sines or two cosines into a product, use the appropriate sum-product formula.

- $\cos\alpha + \cos\beta = 2\cos\left(\frac{\alpha+\beta}{2}\right)\cos\left(\frac{\alpha-\beta}{2}\right)$
- $\cos\alpha - \cos\beta = -2\sin\left(\frac{\alpha+\beta}{2}\right)\sin\left(\frac{\alpha-\beta}{2}\right)$
- $\sin\alpha + \sin\beta = 2\sin\left(\frac{\alpha+\beta}{2}\right)\cos\left(\frac{\alpha-\beta}{2}\right)$
- $\sin\alpha - \sin\beta = 2\cos\left(\frac{\alpha+\beta}{2}\right)\sin\left(\frac{\alpha-\beta}{2}\right)$

**Example 10:** Evaluate using a sum-product formula.

$$\sin\frac{23\pi}{12} + \sin\frac{7\pi}{12}$$

Neither of these angles appear on the unit circle, but notice that both half of their sum and half of their difference are angles which do. This makes the problem a prime candidate for a sum-product formula.

$$\frac{1}{2}\left(\frac{23\pi}{12}+\frac{7\pi}{12}\right)=\frac{1}{2}\left(\frac{30\pi}{12}\right)=\frac{5\pi}{4}$$

$$\frac{1}{2}\left(\frac{23\pi}{12}-\frac{7\pi}{12}\right)=\frac{1}{2}\left(\frac{16\pi}{12}\right)=\frac{2\pi}{3}$$

Plug into the sin α + sin β formula, using the simplified expressions above.

$$\begin{aligned}\sin\frac{23\pi}{12}+\sin\frac{7\pi}{12}&=2\sin\left(\frac{5\pi}{4}\right)\cos\left(\frac{2\pi}{3}\right)\\&=2\left(-\frac{\sqrt{2}}{2}\right)\left(-\frac{1}{2}\right)\\&=\frac{\sqrt{2}}{2}\end{aligned}$$

### Product-sum formulas

These formulas are used to rewrite products of sines and/or cosines into equivalent sums. Like sum-product formulas, these are used very rarely compared to the identities that preceded them, but are invaluable if and when the need arises.

- $\cos\alpha\cos\beta=\dfrac{\cos(\alpha-\beta)+\cos(\alpha+\beta)}{2}$
- $\sin\alpha\sin\beta=\dfrac{\cos(\alpha-\beta)-\cos(\alpha+\beta)}{2}$
- $\cos\alpha\sin\beta=\dfrac{\sin(\alpha+\beta)-\sin(\alpha-\beta)}{2}$
- $\sin\alpha\cos\beta=\dfrac{\sin(\alpha+\beta)+\sin(\alpha-\beta)}{2}$

If you notice that α + β and α – β both equal a unit circle angle, use these formulas to evaluate the expression, much like Example 10.

## Oblique Triangle Laws

In Chapter 5, you used reference angles to calculate trigonometric values of oblique angles in standard position on the coordinate plane. The laws you'll learn in this section apply not to oblique angles but oblique triangles. The angles in these triangles are usually written in degrees, rather than radians, but the statements hold true no matter how the angles are measured.

For the sake of uniformity, these theorems will refer to angles and sides in a standard way. While you cannot assume a given triangle is acute or obtuse, you can assume that the letters representing angles and their opposite sides will match. For example, angle $A$ is opposite side $a$, and angle $C$ is opposite side $c$, as shown in Figure 6-1.

**Figure 6-1** In Law of Sines and Law of Cosines problems, matching letters indicate angles and sides opposite one another.

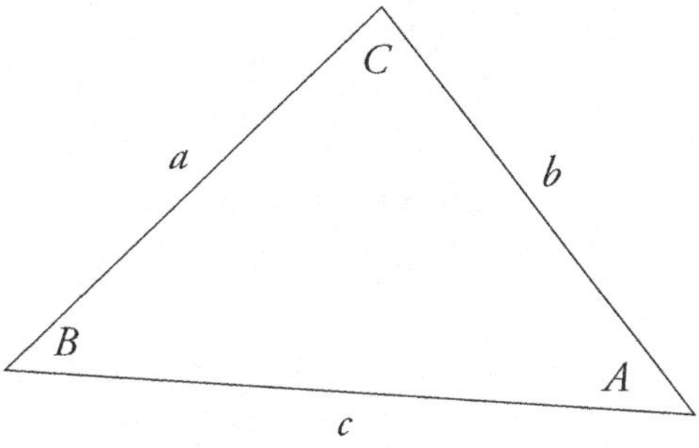


## Law of Sines

According to the **Law of Sines**, the angles of a triangle and the lengths of their opposite sides are all in the same proportion:

$$\frac{\sin A}{a} = \frac{\sin B}{b} = \frac{\sin C}{c}$$

Sometimes, this formula is written as an equality statement of the reciprocals, which is also true.

$$\frac{a}{\sin A} = \frac{b}{\sin B} = \frac{c}{\sin C}$$

The Law of Sines helps you calculate angles and sides of a triangle if you are given either:

- Measures of two angles in the triangle and the length of a side
- Measures of two of the sides in the triangle, and an angle not included (created by) those sides

The Law of Sines will always work if given two angle measurements, but in the case of two sides and a non-included angle, there is a chance you could get more than one possible solution or no solution at all. This weakness will be addressed by the next law, the Law of Cosines.

**Example 11:** Given an oblique triangle $ABC$ with $A = 55°$, $B = 73°$, and $a = 6$, find the length of side $c$.

Since the sum of the angles of a triangle equals 180°, you can find the measure of angle $C$.

$$C = 180° - 55° - 73° = 52°$$

Apply the Law of Sines using $c$, since it's the value you're trying to find, and $a$, since you know both its length and the measure of its opposite angle.

$$\frac{\sin A}{a} = \frac{\sin C}{c}$$

$$\frac{\sin 55^\circ}{6} = \frac{\sin 52^\circ}{c}$$

$$c \cdot \sin 55° = 6 \cdot \sin 52°$$

$$c = \frac{6 \sin 52^\circ}{\sin 55^\circ} \approx 5.772$$

This answer makes sense, because the larger an angle, the larger the side opposite it will be. Since $C$ is slightly smaller than $A$, then $c$ is in turn slightly smaller than $a$.

**Example 12:** Given an oblique triangle $ABC$ with $C = 20°$, $b = 13$, and $c = 5$, find the measure of angle $B$.

Apply the Law of Sines.

$$\frac{\sin B}{b} = \frac{\sin C}{c}$$

$$\frac{\sin B}{13} = \frac{\sin 20^\circ}{5}$$

$$\sin B = \frac{13 \sin 20^\circ}{5}$$

$$\sin B \approx .8892523726$$

Using a calculator, you can calculate

$$\arcsin (.8892523726) \approx 62.78°$$

However, a second quadrant reference angle of 62.78° produces the same sine value, since sine is positive in both the first and second quadrants. Since this angle could be acute (first quadrant) or obtuse (second quadrant), both answers are acceptable; calculate the angle which corresponds to that reference angle:

$$180° - 62.78° = 117.22°$$

There are two possible triangles that could possess the measurements given to you in the problem, so both measurements for $B$ are correct.

## Law of Cosines

If you are not given the information required by the Law of Sines (i.e. a pair of angles or two sides and a non-included angle), the problem may be a candidate for the **Law of Cosines**, which states that the sides and angles of an oblique triangle are related in this manner:

$$a^2 = b^2 + c^2 - 2bc \cos A$$

$$b^2 = a^2 + c^2 - 2ac \cos B$$

$$c^2 = a^2 + b^2 - 2ab \cos C$$

The Law of Cosines is better suited than the Law of Sines to help find missing angles and sides of triangles if given either:

- The lengths of two sides and the measure of their included angle
- The lengths of all three sides of the triangle, but no angle measurements

If a problem asks you to calculate the measures of multiple parts of a triangle, and you must use the Law of Cosines to begin that problem, you need not stick with the Law of Cosines to complete it. Once you have the measurements of an angle and its opposite side, you can default back to the Law of Sines (which is shorter and requires less work).

However, if you are calculating an angle which *might be obtuse*, you must use the Law of Cosines to get the correct angle measurement. Remember, the Law of Sines cannot always differentiate between acute and obtuse angles. However, the Law of Cosines can, since cosine has different signs in quadrants I and II (unlike sine).

**Example 13:** Given an obtuse triangle $ABC$ with $a = 17$, $b = 26$, and $c = 14$, find the measurements of all the angles accurate to the hundredths place.

Since you are given all of the sides' lengths, apply the Law of Cosines. Start by calculating the measure of $B$ since it is opposite the largest side, and therefore possibly obtuse.

$$b^2 = a^2 + c^2 - 2ac \cos B$$

$$26^2 = 17^2 + 14^2 - 2(17)(14) \cos B$$

$$676 = 485 - 476\cos B$$

$$\arccos\left(\frac{-191}{476}\right) = B$$

Evaluate using a calculator.

$$B \approx 113.6570024°$$

Do not round any angle measurements until the end, or you risk compounded inaccuracies. Now use the Law of Sines to calculate one of the remaining angles.

$$\frac{\sin A}{a} = \frac{\sin B}{b}$$

$$\frac{\sin A}{17} = \frac{\sin 113.6570024°}{26}$$

$$\sin A \approx 17 \cdot .0352293837$$

$$A \approx 36.79112242°$$

Since the sum of the angles of a triangle is 180°, calculate $C$.

$$C = 180° - 113.6570024 - 36.79112242 \approx 29.55187518°$$

Therefore, $A = 36.79°$, $B = 113.66°$, and $C = 29.55°$.

## Calculating Triangle Area

The most basic and most widely known formula for the area of a triangle is

$$A = \frac{1}{2}bh$$

where $b$ is the base and $h$ is the height. However, this formula is only useful when the segments representing $b$ and $h$ meet at a right angle. Therefore, if you are trying to calculate the area of an oblique triangle (or a triangle whose height is not easy to find), this formula isn't applicable. This section presents additional formulas that allow you to find a wider variety of triangle areas.

## Given Side-Angle-Side

If you are given the lengths of two sides of a triangle ($x$ and $y$) and the measure of their included angle ($\theta$), then the area of the triangle will be

$$\frac{1}{2}xy\sin\theta$$

**Example 14:** Find the area of the triangle pictured in Figure 6-2.

**Figure 6-2** The diagram for Example 14.

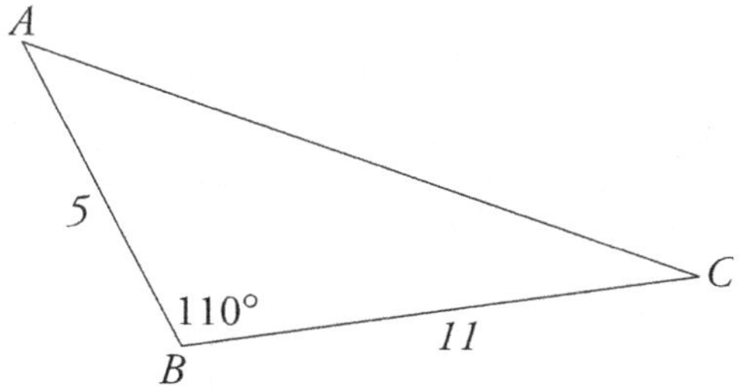


Since the angle 110° is contained by the sides of length 5 and 11 (in other words, you are given side-angle-side), apply the sine area formula.

$$Area = \frac{1}{2} \cdot 5 \cdot 11 \cdot \sin 110°$$
$$Area \approx 25.842 \text{ square units}$$

This formula (like the Law of Sines and the Law of Cosines) will work equally well for angles expressed as either degrees or radians, since an angle has the same sine or cosine value, no matter what units you use to measure it.

### Given side-side-side

If you are given no angle measurements, but do know the lengths of all an oblique triangle's sides, you can apply **Heron's Area Formula** to find the area of the triangle. According to this formula, a triangle with side lengths $a$, $b$, and $c$ will have area

$$\sqrt{s(s-a)(s-b)(s-c)}$$
$$\text{if } s = \frac{a+b+c}{2}.$$

**Example 15:** Calculate the area of the triangle in Figure 6-3.

**Figure 6-3** The triangle described in Example 15.

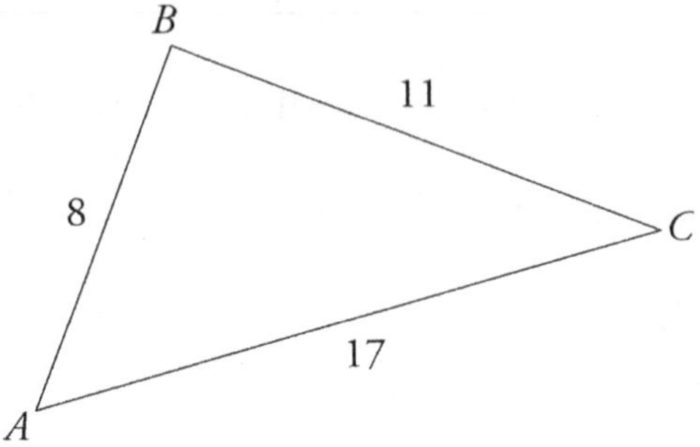


Begin by calculating *s*.

$$s = \frac{8 + 11 + 17}{2}$$
$$s = 18$$

Use this value in Heron's Area Formula.

$$\begin{aligned} Area &= \sqrt{s(s-a)(s-b)(s-c)} \\ &= \sqrt{18(18-8)(18-11)(18-17)} \\ &= \sqrt{1260} \\ &\approx 35.50 \text{ square units} \end{aligned}$$

## Chapter Checkout

### Q&A

1. Prove the identity: $\tan^2\theta = \frac{\sec\theta - \cos\theta}{\cos\theta}$.
2. Solve the trigonometric equation and give all solutions on the interval $[0,2\pi)$: $8\cos(2\theta) - 4\sqrt{3} = 0$.
3. Given an oblique triangle containing sides of length 4 and 7, such that the angle formed by those sides measures 132°, answer the following:
   (a) What is the length of the third side?
   (b) What is the area of the triangle?
4. Calculate the area of a triangle with sides of length 6, 7, and 11.

**Answers:**

**1.**

$$\tan^2\theta = \frac{\sec\theta}{\cos\theta} - \frac{\cos\theta}{\cos\theta}$$

$$\tan^2\theta = \frac{\sec\theta}{\frac{1}{\sec\theta}} - 1$$

$$\tan^2\theta = \sec^2\theta - 1$$

$$1 + \tan^2\theta = \sec^2\theta$$

**2.** $\frac{\pi}{12}, \frac{11\pi}{12}, \frac{13\pi}{12}, \frac{23\pi}{12}$

**3. (a)** $b \approx 10.123$ **(b)** 10.404 **4.** $6\sqrt{10}$

# Chapter 7

# VECTORS AND THE TRIGONOMETRY OF COMPLEX NUMBERS

## Chapter Check-In

- ❑ Interpreting vectors algebraically and geometrically
- ❑ Performing operations on vectors in component form
- ❑ Calculating the dot product
- ❑ Expressing complex numbers trigonometrically
- ❑ Applying DeMoivre's Theorem

This chapter contains trigonometric applications for other fields of mathematic study—vectors and complex numbers—which demonstrates that trigonometry is useful for more than calculating the angles and sides of triangles.

## Vectors in the Coordinate Plane

A **vector** is a quantity that possesses two characteristics: magnitude and direction. *Vectors* are visualized as directed line segments (segments with an arrow head at one end indicating direction) on the coordinate plane, as demonstrated in Figure 7-1.

**Figure 7-1** Vector **v** travels from point $P$ to point $Q$.

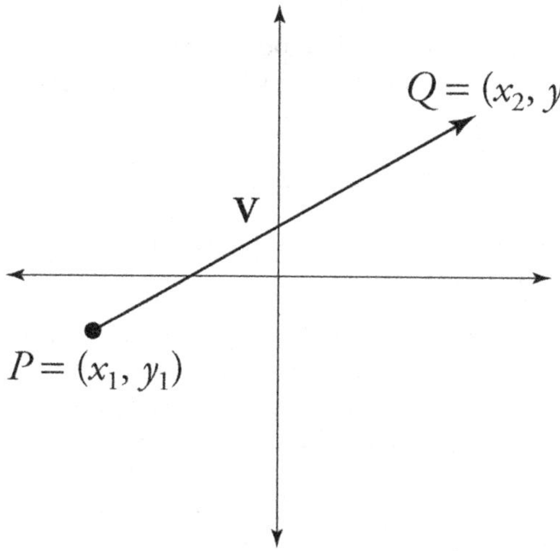


The vector in Figure 7-1 begins at **initial point** $P$ and ends at **terminal point** $Q$. Whereas vectors are usually expressed as variables in boldface, they can also be written in terms of their endpoints. Therefore, the vector in Figure 7-1 could be named either **v** or $\overrightarrow{PQ}$.

## Standard form of a vector

Vector arithmetic is vastly simplified if the vectors involved have initial points which lie on the origin of the coordinate plane. Such vectors are said to be in **standard position.**

Once a vector is in *standard position*, the coordinates of its terminal point, $(x_2, y_2)$, can be written as $<x_2, y_2>$. The change of grouping symbol indicates that the vector is now in **component form**, since $x_2$ and $y_2$ represent the horizontal and vertical components of the vector, respectively. In other words, if vector **m** has *component form* $<a,b>$, then its initial point is the origin, and its terminal point is $(a,b)$.

**Example 1:** Write vector **n**, which has initial point (–3,7) and terminal point (4,–2) in component form.

Subtract the *x*-values of the terminal point from the *x*-values of the initial point to get the *x*-value of component form. Repeat the process for the *y*-values.

$$\mathbf{n} = <4 - (-3), -2 - 7> = <7,-9>$$

Notice that the terminal point of **n** is exactly 7 units to the right of and 9 units down from its initial point.

Once a vector is in component form, you can easily find the length of the vector (called its **magnitude**) using the Pythagorean Theorem, since a right triangle is formed between the origin, the terminal point of the vector, and the point on the x-axis directly below the terminal point. The *magnitude* of a vector **v** is written $\|\mathbf{v}\|$.

$$\text{If } \mathbf{v} = <a,b>, \text{ then } \|\mathbf{v}\| = \sqrt{a^2 + b^2}$$

**Example 2:** Calculate the magnitude of $\mathbf{n} = <7,-9>$.

$$\|\mathbf{n}\| = \sqrt{7^2 + (-9)^2} = \sqrt{49 + 81} = \sqrt{130}$$

## Unit vectors

A vector which has a magnitude of 1 is called a **unit vector**, much like the circle of radius 1 is called the unit circle. You may be asked to find a unit vector in the same direction as a given vector. A vector, **v**, written in component form, can be transformed into a unit vector of the same direction, $\mathbf{v}_0$, by dividing each component of **v** by $\|\mathbf{v}\|$.

**Example 3:** If $\mathbf{v} = <3,-2>$, calculate the unit vector $\mathbf{v}_0$ which shares the same direction as **v**.

First find the magnitude of **v**.

$$\|\mathbf{v}\| = \sqrt{3^2 + (-2)^2} = \sqrt{13}$$

Divide each component of **v** by $\|\mathbf{v}\|$ to find $\mathbf{v}_0$.

$$\mathbf{v}_0 = \left\langle \frac{3}{\sqrt{13}}, -\frac{2}{\sqrt{13}} \right\rangle = \left\langle \frac{3\sqrt{13}}{13}, -\frac{2\sqrt{13}}{13} \right\rangle$$

There are two standard unit vectors, $\mathbf{i} = <1,0>$ and $\mathbf{j} = <0,1>$, which represent horizontal and vertical components, respectively. Any vector $\mathbf{v} = <a, b>$ can be written in terms of these unit vectors

$$\mathbf{v} = a\mathbf{i} + b\mathbf{j}$$

The reverse is also true; $\mathbf{w} = c\mathbf{i} + d\mathbf{j}$ is equivalent to $\mathbf{w} = <c,d>$.

**Example 4:** Vector **v** has initial point $P = (0,4)$ and terminal point $Q = (-3,5)$. Write **v** in terms of standard unit vectors **i** and **j**.

Begin by expressing the vector in component form.

$$\mathbf{v} = <-3 - 0, 5 - 4>$$

$$\mathbf{v} = <-3,1>$$

The horizontal component is the coefficient for **i**, and the vertical component is the coefficient for **j**.

$$\mathbf{v} = -3\mathbf{i} + \mathbf{j}$$

### Basic vector operations

There are four major operations you will perform on vectors. You should understand the following three both algebraically and geometrically. The fourth will be discussed in the next section.

- **Vector addition:** If vector $\mathbf{v} = <a,b>$ and $\mathbf{w} = <c,d>$, then

  $$\mathbf{v} + \mathbf{w} = <a + c, b + d>$$

  To add vectors in the coordinate plane, place the initial point of the second vector on top of the terminal point of the first vector. The vector representing the sum will have the same initial point as the first vector and the same terminal point as the second vector. This is called the **parallelogram law** of vector addition; it's pictured in Figure 7-2.

**Figure 7-2** The parallelogram law for vector addition.

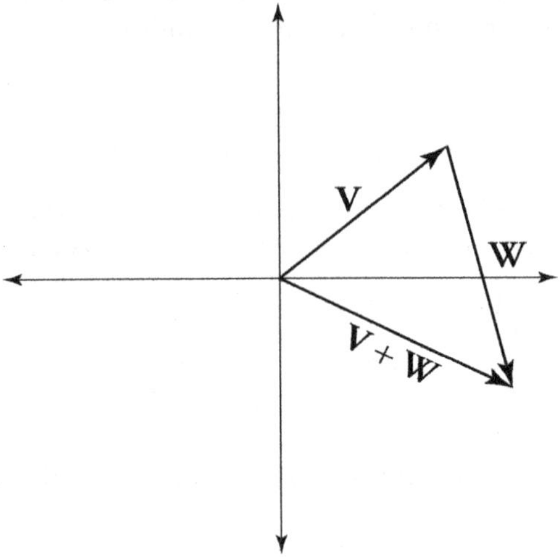


**Example 5:** If **v** = <–3,10> and **w** = <8,–1>, calculate **v** + **w**.

$$\mathbf{v} + \mathbf{w} = <-3 + 8, 10 - 1> = <5,9>$$

- **Scalar multiplication:** When a vector is multiplied by a constant (also called a **scalar**), both the length and the direction of the vector can be changed.

  If **v** = <*a*,*b*> has magnitude *m*, then *c***v** = <*ca*,*cb*> and $||c\mathbf{v}|| = |cm|$

  If $c < 0$, then vector *c***v** will travel in the opposite direction as **v**.

Both possible effects of scalar multiplication are demonstrated geometrically in Figure 7-3.

**Figure 7-3** The effects of scalar multiplication on vector **v**.

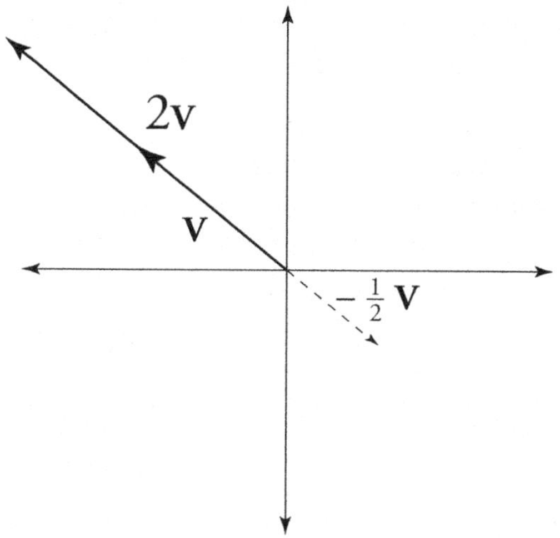


**Example 6:** If **v** = –4**i** + 11**j**, evaluate –9**v**.

Written in component form, **v** = <–4,11>. Multiply each component by –9.

$$-9\mathbf{v} = <-4 \cdot -9, 11 \cdot -9> = <36,-99>$$

- **Vector subtraction:** Any vector subtraction problem can be rewritten as a vector addition problem with a scalar multiple of –1 times the second vector:

  $$\mathbf{v} - \mathbf{w} = \mathbf{v} + (-\mathbf{w})$$

  To visualize a subtraction problem, reverse the direction of the second vector (but leave its length alone). Then, apply the parallelogram law for vector addition and treat it as an addition problem.

**Example 7:** If **v** = <–2,3> and **w** = <1,–6>, calculate **v** – **w**.

$$\mathbf{v} - \mathbf{w} = \mathbf{v} + (-\mathbf{w}) = <-2,3> + <-1,6> = <-3,9>$$

# Dot Products

The **dot product** of two vectors, $\mathbf{v} = <a,b>$ and $\mathbf{w} = <c,d>$, is defined as

$$\mathbf{v} \cdot \mathbf{w} = ac + bd$$

## Properties of the dot product

Notice that the result is a scalar (a number), not a vector, unlike the result of the preceding three vector operations.

The dot product has the following properties:

- $\mathbf{v} \cdot \mathbf{w} = \mathbf{w} \cdot \mathbf{v}$ (the dot product is commutative)
- $\mathbf{u} \cdot (\mathbf{v} + \mathbf{w}) = \mathbf{u} \cdot \mathbf{v} + \mathbf{u} \cdot \mathbf{w}$ (the dot product can be distributed through a vector sum)
- $n(\mathbf{v} \cdot \mathbf{w}) = n\mathbf{v} \cdot \mathbf{w}$ or $\mathbf{v} \cdot n\mathbf{w}$, where $n$ is a scalar (if a dot product is multiplied by a scalar, the expression can be rewritten as a dot product with $n$ as a scalar multiple of exactly one of the vectors)
- $\mathbf{v} \cdot \mathbf{0} = 0$ ($\mathbf{0}$ is known as the **zero vector** and has component form $<0,0>$)
- $\mathbf{v} \cdot \mathbf{v} = ||\mathbf{v}||^2$ (the dot product of a vector with itself is equal to the square of that vector's magnitude)

**Example 8:** Evaluate the following if $\mathbf{v} = <2,-3>$ and $\mathbf{w} = <-8,1>$.

(a) $\mathbf{w} \cdot \mathbf{v}$

$$(-8 \cdot 2) + (1 \cdot -3) = -19$$

(b) $2\mathbf{v} \cdot -3\mathbf{w}$

$$(2 \cdot <2,-3>) \cdot (-3 \cdot <-8,1>)$$

$$<4,-6> \cdot <24,-3>$$

$$96 + 18 = 114$$

(c) $2(\mathbf{v} \cdot \mathbf{w})$

You can either take the dot product first and then multiply by 2 or multiply one of the vectors by 2, and then take the dot product. Either of the techniques will result in the same answer. Below, both techniques are demonstrated.

$$2(\mathbf{v} \cdot \mathbf{w}) = \mathbf{v} \cdot 2\mathbf{w}$$

$$2(<2,-3> \cdot <-8,1>) = <2,-3> \cdot 2<-8,1>$$

$$2(-16 - 3) = <2,-3> \cdot <-16,2>$$

$$2(-19) = -38$$

(d) $<5,-3> \cdot (\mathbf{v} + \mathbf{w})$

Distribute the vector <5,–3> across the vector sum of **v** and **w**.

$$<5,-3> \cdot \mathbf{v} + <5,-3> \cdot \mathbf{w}$$

$$<5,-3> \cdot <2,-3> + <5,-3> \cdot <-8,1>$$

$$19 - 43 = -24$$

(e) $\mathbf{w} \cdot \mathbf{w}$

The below work demonstrates that the dot product of a vector with itself is equal to the square of its magnitude; note that $\|\mathbf{w}\| = \sqrt{65}$.

$$\mathbf{w} \cdot \mathbf{w} = \|\mathbf{w}\|^2$$

$$\langle -8, 1\rangle \cdot \langle -8, 1\rangle = \left(\sqrt{65}\right)^2$$

$$64 + 1 = 65$$

## Measuring angles between vectors

The angle θ ($0 \le \theta \le \pi$) between vectors **v** and **w** in standard position, as shown in Figure 7-4, can be found with this formula:

$$\cos\theta = \frac{\mathbf{v} \cdot \mathbf{w}}{\|\mathbf{v}\| \, \|\mathbf{w}\|}$$

**Example 9:** Calculate the measure of θ, the angle formed by **v** = <7,6> and **w** = <–4,2>.

Substitute the vectors into the angle formula.

$$\cos\theta = \frac{\langle 7, 6\rangle \cdot \langle -4, 2\rangle}{\|\mathbf{v}\|\|\mathbf{w}\|}$$

$$\|\mathbf{v}\| = \sqrt{7^2 + 6^2} = \sqrt{85}$$

$$\|\mathbf{w}\| = \sqrt{(-4)^2 + 2^2} = \sqrt{20} = 2\sqrt{5}$$

$$\cos\theta = \frac{-28 + 12}{\sqrt{85} \cdot 2\sqrt{5}}$$

$$\cos\theta \approx -.388057000058$$

$$\theta \approx 112.83^\circ$$

Note that if the vector names were reversed in this example (**v** = <–4,2> and **w** = <7,6>), the result would still be the same. That's because both the numerator and the denominator of the angle formula contain commutative operations.

**Figure 7-4** If **v** and **w** are in standard position, you can calculate the measure of θ, the angle between them.

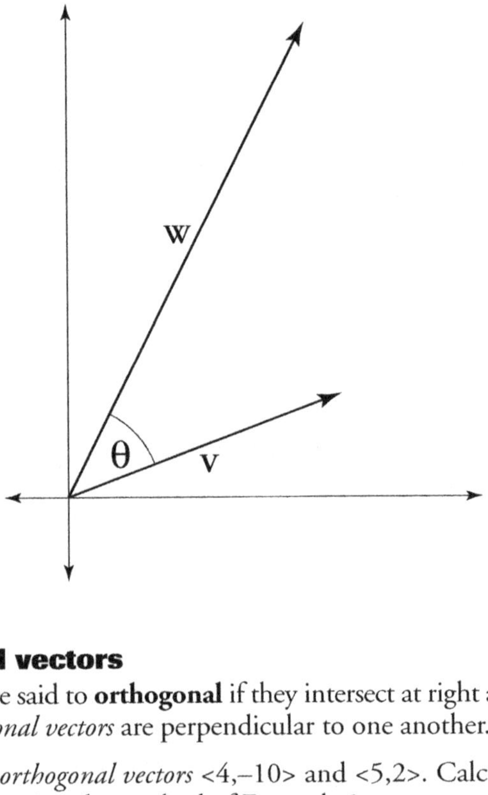


## Orthogonal vectors

Two vectors are said to **orthogonal** if they intersect at right angles. In other words, *orthogonal vectors* are perpendicular to one another.

Consider the *orthogonal vectors* <4,–10> and <5,2>. Calculate the angle between them using the method of Example 9.

$$\cos\theta = \frac{\langle 4, -10\rangle \cdot \langle 5, 2\rangle}{\|\mathbf{v}\|\|\mathbf{w}\|}$$

$$\cos\theta = \frac{20 - 20}{\|\mathbf{v}\|\|\mathbf{w}\|}$$

$$\cos\theta = \frac{0}{\|\mathbf{v}\|\|\mathbf{w}\|}$$

$$\cos\theta = 0$$

$$\theta = \frac{\pi}{2} = 90^\circ$$

From this example, you can see that the magnitudes of the vectors are irrelevant when determining whether or not vectors are orthogonal. In fact, the only condition that non-zero orthogonal vectors **v** and **w** must satisfy is

$$\mathbf{v} \cdot \mathbf{w} = 0$$

Since the dot product is commutative, a result of $\mathbf{w} \cdot \mathbf{v} = 0$ is sufficient as well. Note that although zero vectors are orthogonal to all vectors, they do not exhibit the geometric property of perpendicularity, since zero vectors have zero magnitude.

**Example 10:** Find the value of $a$ which makes $\mathbf{v} = \langle -5,6 \rangle$ and $\mathbf{w} = \langle 4,a \rangle$ orthogonal vectors.

If **v** and **w** are orthogonal, then $\mathbf{v} \cdot \mathbf{w} = 0$.

$$\langle -5,6 \rangle \cdot \langle 4,a \rangle = 0$$

Evaluate the dot product and solve for $a$.

$$-20 + 6a = 0$$

$$6a = 20$$

$$a = \frac{10}{3}$$

## Complex Numbers and Trigonometry

A complex number, as you may recall from Chapter 1, looks like $a + bi$ and contains two elements: a real part, $a$, and an imaginary part, $bi$, where $i = \sqrt{-1}$.

Because it possesses two components, a complex number cannot be graphed on a number line, like a real number can. Instead, complex numbers must be graphed on a coordinate plane, where the horizontal axis represents the real quantity and the vertical axis represents the imaginary quantity, as pictured in Figure 7-5.

**Figure 7-5** Complex numbers graphed in the coordinate plane.

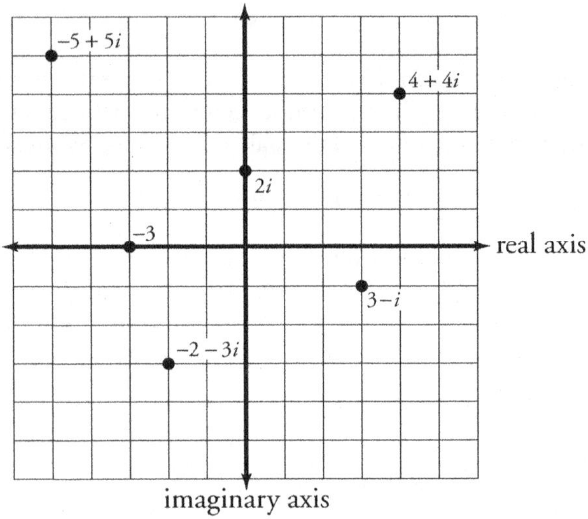


## Basic operations with complex numbers

Adding and subtracting complex numbers is simple. Just combine the real parts and do the same with the imaginary parts, as though you were adding like terms of different polynomials.

$$(3 + 4i) + (-2 + 6i) = (3 - 2) + (4i + 6i) = 1 + 10i$$

$$(6 - 3i) - (-10 + 5i) = (6 - 3i) + (10 - 5i) = 16 - 8i$$

Multiplication and division of complex numbers is only slightly more difficult.

**Example 11:** Find the product: $(1 + 3i)(-2 - 7i)$.

Use the FOIL method, as though you were multiplying linear binomials.

$$-2 - 7i - 6i - 21i^2$$

Combine the imaginary terms.

$$-2 - 13i - 21i^2$$

Since $i = \sqrt{-1}$, then $i^2 = \left(\sqrt{-1}\right)^2 = -1$.

$$-2 - 13i - 21(-1)$$

$$-2 - 13i + 21$$

$$19 - 13i$$

The product of two complex numbers will simplify to a complex number itself, once the $i^2$ term is simplified.

The **conjugate** of a complex number $(a \pm bi)$ is equal to $a \mp bi$; change the middle sign to its opposite.

**Example 12:** Find the quotient: $\frac{2i}{4 - 3i}$.

Multiply both the numerator and denominator of the quotient by the conjugate of the denominator. This will eliminate imaginary numbers from the denominator. Simplify the result.

$$\frac{2i}{4 - 3i} \cdot \frac{4 + 3i}{4 + 3i}$$

$$\frac{8i + 6i^2}{16 - 9i^2}$$

$$\frac{8i - 6}{16 + 9}$$

$$\frac{8i}{25} - \frac{6}{25} = -\frac{6}{25} + \frac{8i}{25}$$

Just like multiplication, the result of dividing two complex numbers is itself a complex number.

## Trigonometric form of a complex number

The complex number $c = a + bi$ can be written in trigonometric form

$$c = r(\cos\theta + i\sin\theta)$$

where the variables are defined as follows:

- $r = \sqrt{a^2 + b^2}$ (the distance from the origin to the point on the coordinate place representing the graph if $c$); $r$ is called either the **modulus** or the **absolute value** of $c$.
- $\theta$ is the angle measured from the positive $x$-axis to the segment joining the origin and $c$; $\theta$ is called the **argument** of $c$. In order to determine the value of $\theta$, notice that $\tan\theta = \frac{b}{a}$.

A graphical representation of each of these variables is given in Figure 7-6.

**Figure 7-6** The variables $a$, $b$, $r$, and $\theta$ will help you put the complex number $c = a + bi$ into trigonometric form.

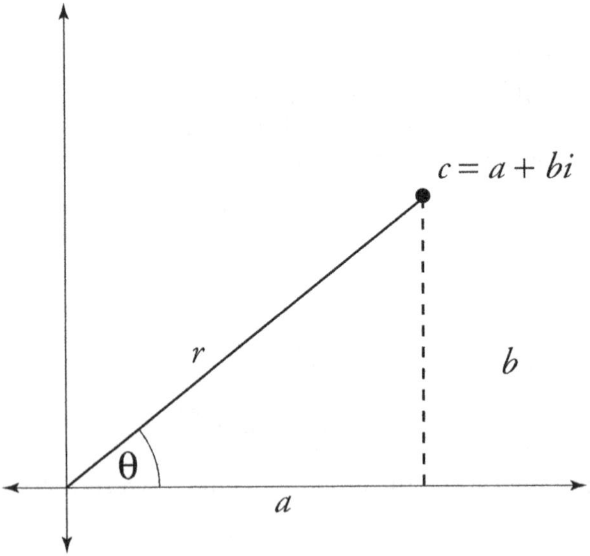


To put a complex number in trigonometric form, follow these steps:

1. Calculate the modulus of $c$.
2. Calculate $\tan\theta$.

3. Determine θ based upon its tangent value and what quadrant it lies in.
4. Substitute $r$ and θ into $c = r(\cos\theta + i\sin\theta)$.

**Example 13:** Write the complex number $c = -3 - \sqrt{3}\,i$ in trigonometric form.

First, find the modulus of $c$.

$$\begin{aligned} r &= \sqrt{(-3)^2 + \left(-\sqrt{3}\right)^2} \\ &= \sqrt{9+3} \\ &= 2\sqrt{3} \end{aligned}$$

Evaluate tan θ.

$$\tan\theta = \frac{-\sqrt{3}}{-3} = \frac{3^{1/2}}{3^1} = 3^{(1/2)-1} = \frac{1}{\sqrt{3}}$$

Since the graph of $c$ lies in the third quadrant (because both its real and imaginary components are negative), chose the θ from the unit circle in that quadrant which has the given tangent value. It is easier to spot θ if you multiply the numerator and denominator of tan θ by $\frac{1}{2}$.

$$\tan\theta = \frac{\sin\theta}{\cos\theta} = \frac{1}{\sqrt{3}} \cdot \frac{\frac{1}{2}}{\frac{1}{2}} = \frac{\frac{1}{2}}{\frac{\sqrt{3}}{2}}$$

$$\theta = \frac{7\pi}{6}$$

Plug θ and $r$ into the formula for trigonometric form.

$$c = r(\cos\theta + i\sin\theta)$$

$$c = 2\sqrt{3}\left(\cos\frac{7\pi}{6} + i\sin\frac{7\pi}{6}\right)$$

Check your answer by evaluating the trigonometric expressions and distributing $2\sqrt{3}$; you'll get $c = -3 - i\sqrt{3}$.

### Multiplying and dividing complex numbers

If $c_1 = r_1(\cos\theta_1 + i\sin\theta_1)$ and $c_2 = r_2(\cos\theta_2 + i\sin\theta_2)$ are complex numbers in trigonometric form, then the product and quotient of $c_1$ and $c_2$ are defined as follows:

$$c_1 \cdot c_2 = r_1 r_2 \left[\cos(\theta_1 + \theta_2) + i\sin(\theta_1 + \theta_2)\right]$$

$$\frac{c_1}{c_2} = \frac{r_1}{r_2}\left[\cos(\theta_1 - \theta_2) + i\sin(\theta_1 - \theta_2)\right]$$

**Example 14:** Evaluate $\frac{c_1}{c_2}$ if

$$c_1 = 3\left(\cos\frac{\pi}{2} + i\sin\frac{\pi}{2}\right) \text{ and } c_2 = 6\left(\cos\frac{\pi}{6} + i\sin\frac{\pi}{6}\right)$$

Apply the quotient formula.

$$\frac{c_1}{c_2} = \frac{3}{6}\left[\cos\left(\frac{\pi}{2} - \frac{\pi}{6}\right) + i\sin\left(\frac{\pi}{2} - \frac{\pi}{6}\right)\right]$$

$$= \frac{1}{2}\left(\cos\frac{\pi}{3} + i\sin\frac{\pi}{3}\right)$$

$$= \frac{1}{2}\left(\frac{1}{2} + i \cdot \frac{\sqrt{3}}{2}\right)$$

$$= \frac{1 + i\sqrt{3}}{4} = \frac{1}{4} + \frac{\sqrt{3}}{4}i$$

## Roots and Powers of Complex Numbers

You may wonder why people use the trigonometric form of a complex number at all, when multiplying and dividing complex numbers in such a form is so much more difficult than simply performing the operations with the complex numbers in form $c = a + bi$.

It turns out that raising complex numbers to exponential powers and finding the roots (like square roots and cube roots, not $x$-intercepts) of complex numbers is a bit easier when those numbers are written in trigonometric form.

### DeMoivre's Theorem

Consider the product of a complex number $c$, written in trigonometric form, with itself.

$$c = r(\cos\theta + i\sin\theta)$$

$$c^2 = [r(\cos\theta + i\sin\theta)]^2$$

Apply the multiplication formula for complex numbers, noting that in this example, $\theta_1 = \theta_2$.

$$c^2 = r \cdot r\left[\cos(\theta+\theta) + i\sin(\theta+\theta)\right]^2$$
$$= r^2(\cos 2\theta + i\sin 2\theta)$$

Using the same process, you can find this result:

$$c^3 = r^3(\cos 3\theta + i\sin 3\theta)$$

This conclusion is generalized for any power in **DeMoivre's Theorem:** If $c = r(\cos\theta + i\sin\theta)$ is a complex number and $n$ is a nonnegative integer, then

$$c^n = r^n(\cos n\theta + i\sin n\theta)$$

**Example 15:** If $c = 2 - 2i$, evaluate $c^5$.

Begin by putting $c$ in trigonometric form. Since $r = 2\sqrt{2}$ and $\theta = \frac{7\pi}{4}$, you get

$$c = 2\sqrt{2}\left(\cos\frac{7\pi}{4} + i\sin\frac{7\pi}{4}\right)$$

Apply DeMoivre's Theorem to calculate $c^5$.

$$c^5 = \left(2\sqrt{2}\right)^5\left(\cos\frac{35\pi}{4} + i\sin\frac{35\pi}{4}\right)$$

Find a coterminal angle for $\frac{35\pi}{4}$ on the unit circle.

$$c^5 = 128\sqrt{2}\left(\cos\frac{3\pi}{4} + i\sin\frac{3\pi}{4}\right)$$

### Calculating *n*th roots of complex numbers

If $w_k$ is an ***n*th root** of the complex number $c = a + bi$, then

$$(w_k)^n = c, \text{ if } n \text{ is a positive integer}$$

This is not dissimilar from the roots discussed in Chapter 1. For example, you know that $x = 4$ is a square root of 16, since

$$(4)^2 = 16$$

If $n$ is a positive integer, then every complex number $c = r(\cos\theta + i\sin\theta)$ has *exactly n* different roots $(w_0, w_1, w_2, \ldots, w_{n-1})$ given by the formula

$$w_k = \sqrt[n]{r}\left(\cos\frac{\theta + 2\pi k}{n} + i\sin\frac{\theta + 2\pi k}{n}\right), \text{ where } k = 0, 1, 2, \ldots, n-1$$

**Example 16:** Find all cube roots (third roots) of $c = 6(\cos\frac{7\pi}{6} + i\sin\frac{7\pi}{6})$.

You're finding third roots, so let $n = 3$. Let $k$ take values from 0 to 1 less than $n$; therefore $k$ will first be equal to 0, then 1, then 2.

$$(k=0):\quad w_0 = \sqrt[3]{6}\left(\cos\frac{7\pi}{18} + i\sin\frac{7\pi}{18}\right)$$

$$(k=1):\quad w_1 = \sqrt[3]{6}\left(\cos\frac{19\pi}{18} + i\sin\frac{19\pi}{18}\right)$$

$$(k=2):\quad w_2 = \sqrt[3]{6}\left(\cos\frac{31\pi}{18} + i\sin\frac{31\pi}{18}\right)$$

## Chapter Checkout

### Q&A

**1.** If **v** has initial point $P = (3,14)$ and terminal point $(7,9)$,

(a) Write **v** in component form.
(b) Calculate $||\mathbf{v}||$.
(c) Write **v** in terms of standard unit vectors **i** and **j**.
(d) Find the unit vector $\mathbf{v}_0$ in component form that has the same direction as **v**.

**2.** If $\mathbf{v} = <-2,3>$ and $\mathbf{w} = <-1,-8>$, calculate the following:

(a) $\mathbf{v} + \mathbf{w}$
(b) $2\mathbf{v} - 3\mathbf{w}$
(c) $\mathbf{v} \cdot \mathbf{w}$
(d) $\theta°$, if $\theta$ is the obtuse angle formed by **v** and **w**

**3.** Calculate the quotient: $\frac{-2+3i}{-1+4i}$.

**4.** Given complex number $c = -1 + \sqrt{3}\,i$,

(a) Write $c$ in trigonometric form.
(b) Calculate $c^3$ using DeMoivre's Theorem.

**Answers: 1. (a)** $<4,-5>$ **(b)** $\sqrt{41}$ **(c)** $\mathbf{v} = 4\mathbf{i} - 5\mathbf{j}$ **(d)** $\left\langle\frac{4}{\sqrt{41}}, -\frac{5}{\sqrt{41}}\right\rangle = \left\langle\frac{4\sqrt{41}}{41}, -\frac{5\sqrt{41}}{41}\right\rangle$ **2. (a)** $<-3,-5>$ **(b)** $<-1,30>$ **(c)** $-22$ **(d)** 139.185° **3.** $\frac{14}{17} + \frac{5}{17}i$ **4. (a)** $c = 2(\cos\frac{2\pi}{3} + i\sin\frac{2\pi}{3})$ **(b)** 8

# Chapter 8
# ANALYTIC GEOMETRY

## Chapter Check-In

- ❑ Expressing equations of conics in standard form
- ❑ Identifying centers, foci, vertices, and asymptotes of conics
- ❑ Understanding parametric equations
- ❑ Transforming coordinates between rectangular and polar form
- ❑ Graphing polar and parametric equations

This chapter presents an in-depth study of non-linear graphs in the coordinate plane. It begins with an analytic (algebraic) interpretation of the four types of conic sections, which were heretofore understood in terms of their geometric definitions. Late in the chapter, you'll find a discussion on different techniques for representing graphs in the coordinate plane.

## Conic Sections

The set of four geometric figures (circles, parabolas, ellipses and hyperbolas) together referred to as the **conic sections** are the cornerstone of modern analytic geometry. They are called the *conic sections* because their shapes can be generated by intersecting a plane with a **double-napped cone**. (A *double-napped cone* is created by joining two right circular cones together at their vertices such that their bases are parallel.) Figure 8-1 illustrates where each conic section can be found on the cone.

Your primary jobs when studying conic sections are:

- Finding an equation of a conic section given its graph or description
- Expressing a conic equation in standard form
- Graphing a conic section given its equation in standard form.

Though all conics have things in common (for example, they all contain at least one perfect square binomial when in standard form), they are different enough to warrant a separate discussion of each.

**Figure 8-1** The four types of conic sections.

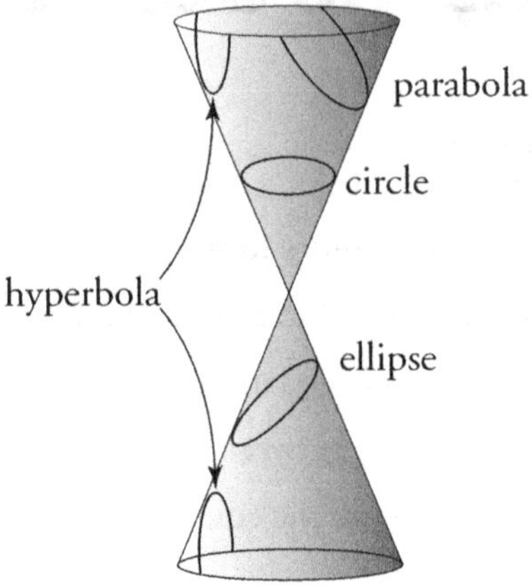


## Circles

A **circle** is a set of coplanar points which are equidistant from a fixed point called the **center**; the distance between each point on the circle and the circle's *center* is called the **radius**. The standard form for the generic circle pictured in Figure 8-2 is

$$(x-h)^2+(y-k)^2=r^2$$

where $(h,k)$ is the center and $r$ is the radius.

**Example 1:** Identify the center and radius of the circle, and use that information to sketch the circle's graph.

$$(x-3)^2+(y+1)^2=16$$

The coordinates for the center are the opposite of the numbers within the squared quantities: (3, –1). The radius is equal to the square root of the constant on the right side of the equation: 4.

To graph the equation, plot the center, and then count four units to the right, units to the left, four units above, and four units below the center.

The circle will pass through those points (as well as all of the other infinitely many points exactly 4 units away from the center) since the radius of the circle is 4 (see Figure 8-3).

**Figure 8-2** The graph of a circle with center $(h,k)$ and radius $r$.

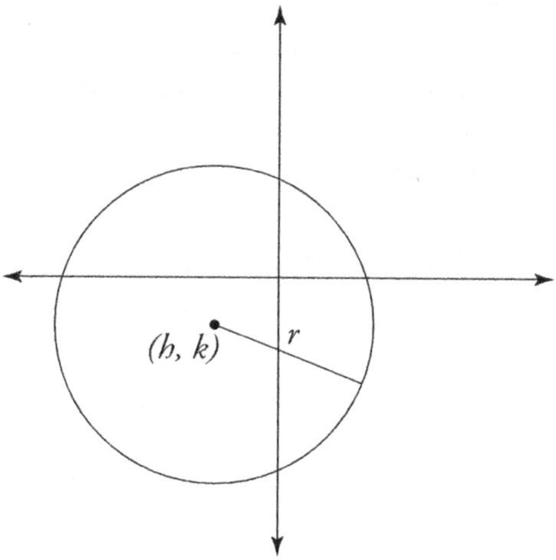


**Figure 8-3** The graph of the circle in Example 1.

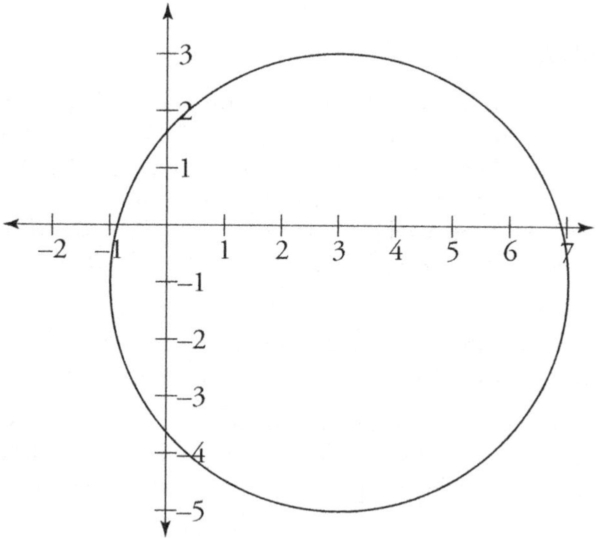

**Example 2:** Find the center and radius of the circle.

$$2x^2 - 8x + 2y^2 - 4y - 62 = 0$$

Your objective will be to rewrite this equation in the standard form of a circle; since that form contains perfect squares, you'll need to complete the square twice in this problem, once for the *x*'s and once for the *y*'s.

Begin by dividing the entire equation by the shared coefficient of the squared terms: 2. Remember, you cannot complete the square unless the coefficients of the squared terms are 1.

$$x^2 - 4x + y^2 - 2y - 31 = 0$$

Move the constant to the right side of the equation.

$$x^2 - 4x + y^2 - 2y = 31$$

Complete the square for both the *x*-terms and the *y*-terms.

$$x^2 - 4x + \mathbf{4} + y^2 - 2y + \mathbf{1} = 31 + \mathbf{4} + \mathbf{1}$$

Factor the left side of the equation.

$$(x - 2)^2 + (y - 1)^2 = 36$$

The center of the circle is (2,1), and the radius is 6.

## Parabolas

A **parabola** is a collection of coplanar points equidistant from a fixed point (called the **focus**) and a fixed line (called the **directrix**) that doesn't contain the *focus*. The point at which the direction of the parabola changes is referred to as the **vertex**, and the line passing through the vertex about which the graph of the parabola is symmetric is called the **axis of symmetry**. Figure 8-4 shows the graphs of two parabolas and all of their corresponding parts.

Notice, as demonstrated by Figure 8-4, that the vertex $V$ is equidistant between the focus and the directrix, along the axis of symmetry.

There are two standard forms for a parabola, just as there are two kinds of parabolas presented in Figure 8-4. You can determine which form to use based upon what variable is squared in the original equation.

**Figure 8-4** The graphs of parabolas with focus point *F* and vertex *V*; *c* is the distance from the vertex to both the focus and the directrix.

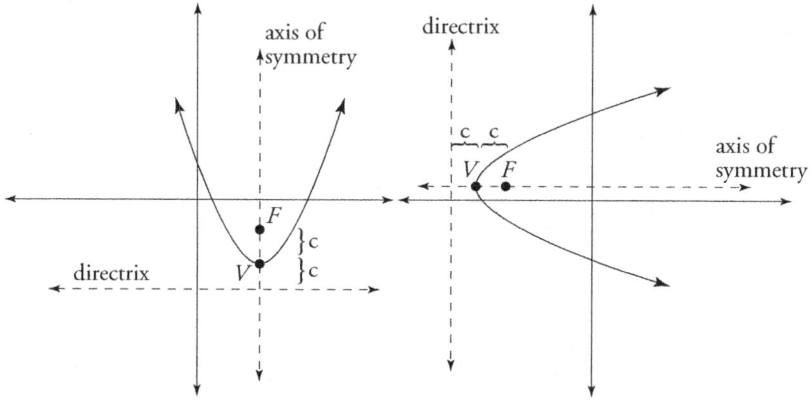


**Standard form 1: For quadratics containing an $x^2$ term**

$$y = a(x - h)^2 + k, \text{ such that } a = \frac{1}{4c}$$

A graph in this form has the following characteristics:

- Opens upward if $a > 0$; opens downward if $a < 0$
- Vertex = $(h,k)$
- Axis of symmetry has equation $x = h$
- Focus = $(h,k \pm c)$, depending upon the direction of the parabola
- Directrix has equation $y = k \pm c$, depending upon the direction of the parabola

Don't get confused by the "±" signs in the formulas both in this and the next standard form. If you draw a diagram, it's easy to tell which way to go from the vertex to get to either the directrix or the focus. Just remember that the parabola will always open in the direction of the focus and away from the directrix.

**Standard form 2: For quadratics containing a $y^2$ term.**

$$x = a(y - k)^2 + h, \text{ such that } a = \frac{1}{4c}$$

Note that both the $x$ and $y$ variables change places, as do the $h$ and $k$ variables when compared with the other standard form. In this version of a parabolic equation, the following characteristics apply:

- Open rightward if $a > 0$; opens leftward if $a < 0$
- Vertex = $(h,k)$, just like in the previous standard form
- Axis of symmetry has equation $y = k$
- Focus = $(h \pm c,k)$, depending upon the direction of the parabola
- Directrix has equation $x = h \pm c$, depending upon the direction of the parabola

**Example 3:** If a parabola has focus (2,–3) and directrix $x = 6$, write the equation of that parabola in standard form.

Since the directrix equation has form "$x =$", then this parabola requires the second standard form equation. (Notice that the other standard form corresponds to parabolas with directrix equations in the form of "$y =$".)

The axis of symmetry is the horizontal line through the focus, so it must have equation $y = -3$. As demonstrated in Figure 8-5, the vertex is the point on the axis of symmetry that's equidistant from both the focus and the directrix, so it must have coordinates (4,–3); $c$ is that equivalent distance, so $c = 2$.

**Figure 8-5** If you plot the focus and directrix, the vertex is the point on the axis of symmetry that falls right in the middle.

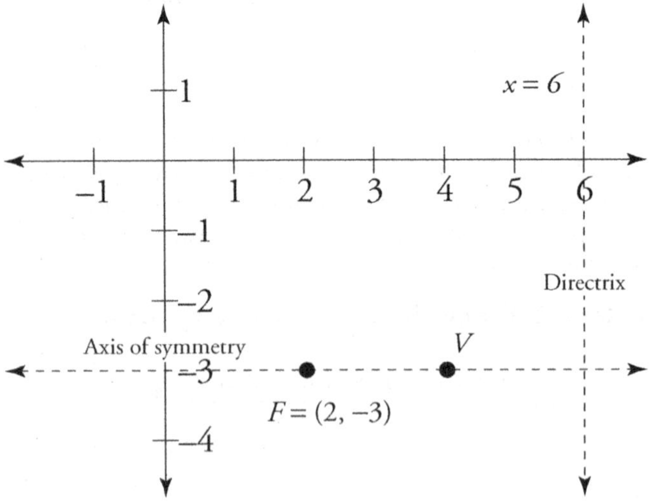

Finally, since a parabola always opens toward its focus and away from its directrix, this parabola must open leftward, making *a* negative. To write the equation, begin by calculating *a*.

$$a = -\frac{1}{4c} = -\frac{1}{4(2)} = -\frac{1}{8}$$

Now substitute the correct values into the second standard form for a parabola.

$$x = a(y - k)^2 + h$$

$$x = -\tfrac{1}{8}\left(y - (-3)\right)^2 + (4)$$

$$x = -\tfrac{1}{8}\left(y + 3\right)^2 + 4$$

**Example 4:** Find the focus, directrix, axis of symmetry, and vertex of the parabola and sketch its graph.

$$y = 2x^2 - 8x + 7$$

Since this equation contains an $x^2$ term, you apply the first standard form equation for a parabola. You'll need to complete the square, so factor a 2 out of both the $x^2$ and $x$ terms.

$$y = 2(x^2 - 4x) + 7$$

Complete the square.

$$y + \mathbf{8} = 2(x^2 - 4x + \mathbf{4}) + 7$$

Even though it looks like you're adding 4 on the right side, that 4 is in a set of parentheses which is multiplied by 2, so you're actually adding $2 \cdot 4 = 8$ to the right side. To keep the equation balanced you must also add 8 to the left side.

Factor and simplify to get the standard form of the parabola.

$$y = 2(x - 2)^2 - 1$$

Figure 8-6 shows the results. The best (and quickest way) to graph parabolas accurately is still to plot points by plugging in a handful of values for *x*. (Knowing the focus and directrix don't really help you plot an accurate graph by hand.) However, if you (and your instructor) are satisfied with a rough sketch, notice that you can graph this parabola using transformations (as discussed in Chapter 2).

**Figure 8-6** The graph of the parabola in Example 4.

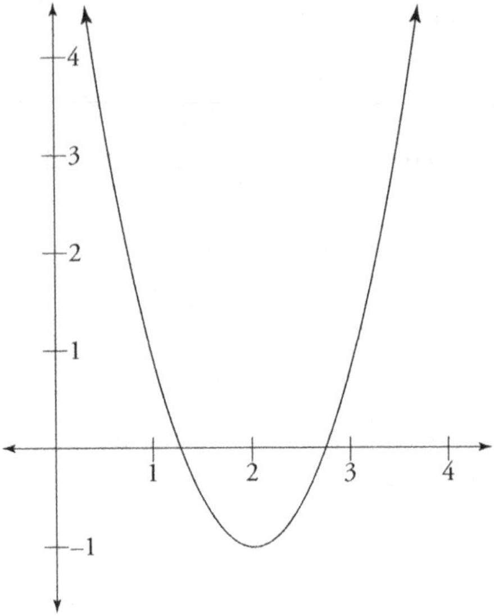


Based on this graph, you can tell that the focus is above the vertex and the directrix is below it (since the graph always opens in the direction of the focus). From the standard form of the equation, you can determine that the vertex is (2,–1) and $a = 2$ (don't forget to take the opposite sign for the quantity within the squared parentheses). Therefore, the axis of symmetry is $x = 2$.

Use $a$ to calculate $c$.

$$2 = \frac{1}{4c}$$

$$8c = 1$$

$$c = \frac{1}{8}$$

Use $c$ to determine the focus and directrix.

$$\text{Focus} = (2, -1 + \tfrac{1}{8}) = (2, -\tfrac{7}{8})$$

$$\text{Directrix: } y = -1 - \frac{1}{8} = \frac{-9}{8}$$

## Ellipses

An **ellipse** is the set of coplanar points such that the sum of the distances from each point to two distinct coplanar points (called the **foci**) is constant. (Note that *foci* is just the plural form of the word *focus*. Therefore, ellipses are defined based on two focus points, as opposed to parabolas, which were only defined based on one.)

The line segment passing through the *foci* is called the **major axis**; its endpoints lie on the ellipse and are called the **vertices**. The midpoint of the major axis is called the **center** of the ellipse.

Every ellipse also possesses a **minor axis**, perpendicular to the *major axis*, which passes through the *center* and has endpoints on the ellipse as well. (Note that the *minor axis* is always shorter than the *major axis*, hence the names.) The parts of an ellipse are illustrated in Figure 8-7.

**Figure 8-7** Two ellipses, one with a horizontal major axis and one with a vertical major axis.

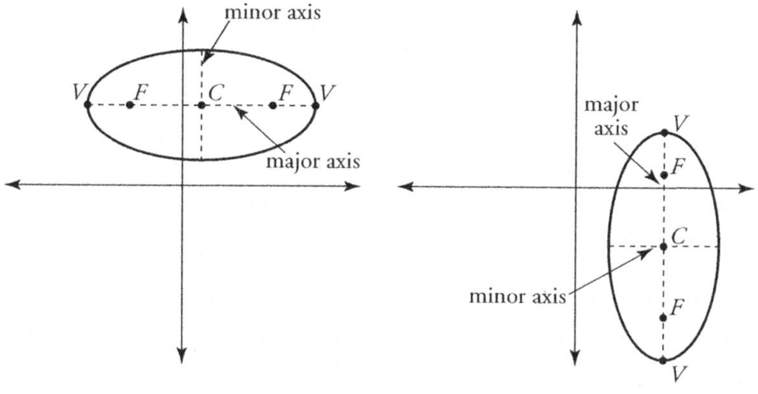


### Standard form

According to mathematical convention, the length of the major axis is $2a$ units. Thus, the distance from the center of the ellipse to each of its vertices is $a$ units. Similarly, the length of the minor axis is $2b$. The variable $c$ is used to measure the distance from the center of the ellipse to each of the foci. See Figure 8-8.

**Figure 8-8** A visual representation of the variables *a*, *b*, and *c* in the standard form of an ellipse.

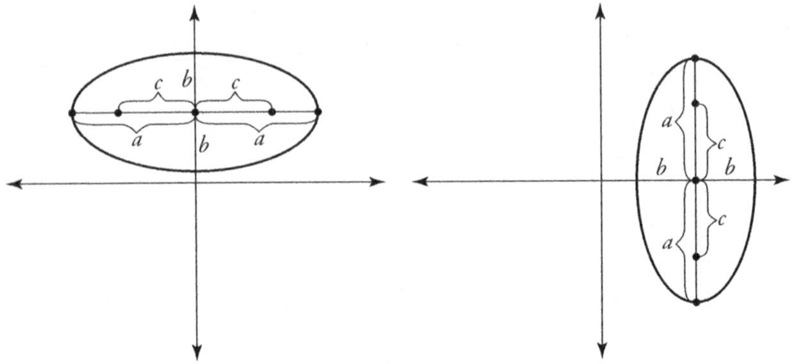


The standard from of an ellipse (with a horizontal major axis) is

$$\frac{(x-h)^2}{a^2}+\frac{(y-k)^2}{b^2}=1$$

If, instead, the major axis is vertical, then the standard form is

$$\frac{(x-h)^2}{b^2}+\frac{(y-k)^2}{a^2}=1$$

In both forms, $(h,k)$ is the center of the ellipse and $c=\sqrt{a^2-b^2}$. As you can see, the only difference between the two forms is the placement of $a^2$ and $b^2$. In each case, the $a^2$ is placed beneath the variable whose axis is parallel to the major axis. For example, if the major axis is horizontal, then $a^2$ is placed beneath the *x* variable, since the *x*-axis is horizontal as well.

**Example 5:** Write the equation of the ellipse whose vertices are (–3,4) and (7,4), with foci (–1,4) and (5,4), in standard form.

The center of the ellipse is the midpoint of both the major axis and the segment whose endpoints are the foci: (2,4). So, $h=2$ and $k=4$. Since the distance between the center and either one of the foci is 3, $c=3$. In addition, the distance from the center to one of the vertices is 5, so $a=5$. Use *a* and *c* to calculate *b*.

$$c=\sqrt{a^2-b^2}$$
$$3=\sqrt{5^2-b^2}$$
$$9=25-b^2$$
$$b^2=16$$
$$b=4$$

Since the line containing the foci and vertices is horizontal, use the standard form for an ellipse with a horizontal major axis.

$$\frac{(x-2)^2}{5^2}+\frac{(y-4)^2}{4^2}=1$$
$$\frac{(x-2)^2}{25}+\frac{(y-4)^2}{16}=1$$

**Example 6:** Put the equation of the ellipse in standard form, and find the coordinates of its center, vertices, and foci. Use that information to sketch its graph.

$$16x^2-128x+y^2+10y=-217$$

Factor 16 out of the $x$-terms so you can complete the square twice, just as you did with circles.

$$16(x^2-8x)+y^2+10y=-217$$
$$16(x^2-8x+\mathbf{16})+y^2+10y+\mathbf{25}=-217+\mathbf{256}+\mathbf{25}$$

Remember, you're not adding 16 and 25 to both sides, you're adding $16\cdot 16=256$ and 25 to both sides. Now simplify and factor the perfect squares.

$$16(x-4)^2+(y+5)^2=64$$

Both standard forms of an ellipse must be set equal to 1, so divide everything by 64 to cancel out the constant.

$$\frac{(x-4)^2}{4}+\frac{(y+5)^2}{64}=1$$

The major axis must be vertical, the center is $(4,-5)$, $a=8$, $b=2$, and $c=\sqrt{64-4}=\sqrt{60}=2\sqrt{15}$. Calculate the coordinates specified by the problem.

$$\text{Vertices: } (4, -5 \pm 8) = (4,3) \text{ and } (4,-13)$$

$$\text{Foci: } \left(4, -5 \pm 2\sqrt{15}\right) = \left(4, -5 + 2\sqrt{15}\right) \text{ and } \left(4, -5 - 2\sqrt{15}\right)$$

To graph the ellipse, plot the center, then count 8 units above and below it to graph the vertices. Since $b = 2$, also plot the points which are 2 units to the left and right of the center. From these 4 points, sketch a rounded, elliptical graph, like in Figure 8-9.

**Figure 8-9** The graph of the ellipse in Example 6.

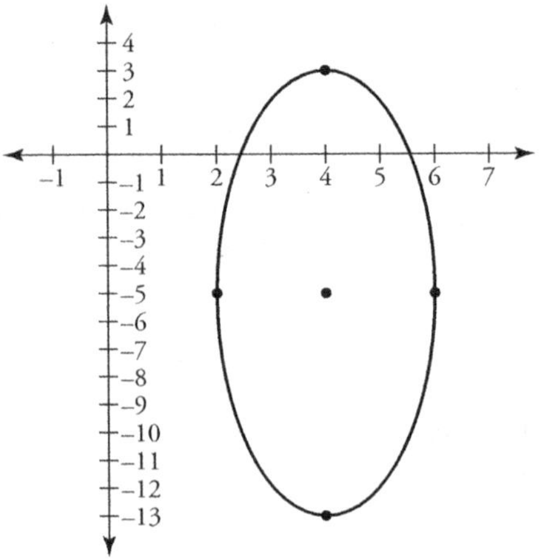


## Eccentricity

The smaller the distance between the foci of an ellipse, the more the ellipse will resemble a circle. In fact, a circle is equivalent to an ellipse whose foci overlap. Generally speaking, an ellipse will fall somewhere within the spectrum ranging from nearly circular to very oval in shape.

The **eccentricity** of an ellipse is a numerical measurement that describes the shape of its graph; its value is

$$e = \frac{c}{a}$$

Since $c$ and $a$ are both always positive and $c$ will always be less than $a$, you know that the *eccentricity* of every ellipse will fall somewhere on the interval (0,1).

The further apart the foci, the greater $c$ is (resulting in a value of $e$ closer to 1), indicating an ellipse which is more oval than circle. If, however, the foci are close together, $c$ will be very small, and $e$ will have a value much closer to 0, indicating that the ellipse is more circle than oval. See Figure 8-10.

**Figure 8-10** The larger the eccentricity of an ellipse, the more it resembles an oval; the closer $e$ is to 0, the more the ellipse resembles a circle.

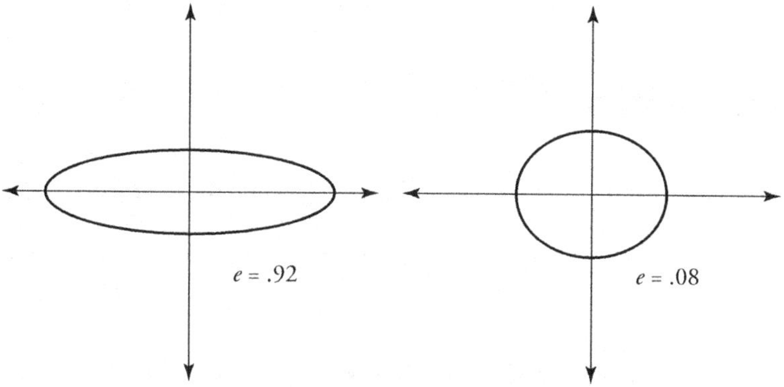


**Example 7:** Determine the eccentricity of the ellipse.

$$x^2 - 2x + 3y^2 - 18y + 16 = 0$$

Put the ellipse in standard form by completing the square like in Example 6.

$$x^2 - 2x + 3(y^2 - 6y) = -16$$

$$x^2 - 2x + \mathbf{1} + 3(y^2 - 6y + \mathbf{9}) = -16 + \mathbf{1} + \mathbf{27}$$

$$(x-1)^2 + 3(y-3)^2 = 12$$

$$\frac{(x-1)^2}{12} + \frac{(y-3)^2}{4} = 1$$

Since $a^2 = 12$ and $b^2 = 4$, you can evaluate $c$.

$$c = \sqrt{12-4} = \sqrt{8} = 2\sqrt{2}$$

Apply the eccentricity formula.

$$\begin{aligned} e &= \frac{c}{a} \\ &= \frac{\sqrt{8}}{\sqrt{12}} \\ &\approx .8165 \end{aligned}$$

According to its eccentricity value, the graph of this ellipse is much more oval than circle.

## Hyperbolas

A **hyperbola** is a set of points such that the difference of the distances from each point to two distinct, fixed points (called the **foci**) is a positive constant. The graph of a *hyperbola* differs from all other conic sections in that its graph is made up of two distinct **branches** that open in opposite directions.

Like an ellipse, a hyperbola is formed by two axes. The line segment passing through the *foci* is called the **transverse axis**, and its endpoints (called the **vertices**) lie on the hyperbola. A hyperbola with a horizontal *transverse axis* opens left and right, whereas one with a vertical *transverse axis* opens up and down. The midpoint of the segment connecting the foci is called the **center** of the hyperbola.

The **conjugate axis** of a hyperbola is perpendicular to the transverse axis at the hyperbola's *center*. The asymptotes of the hyperbola can be used to help visualize the conjugate axis, since its length is not given by any points actually on the hyperbola.

Imagine a rectangle exists nestled between the branches of the hyperbola, such that one pair of its parallel sides pass through the vertices. In addition, the rectangle should be drawn such that the asymptotes of the hyperbola pass exactly through its corners, like in Figure 8-11. The conjugate axis is the segment, perpendicular to the transverse axis, whose end points fall on the rectangle.

**Figure 8-11** The graphs of two hyperbolas with vertices *V*, foci *F*, and center *C*.

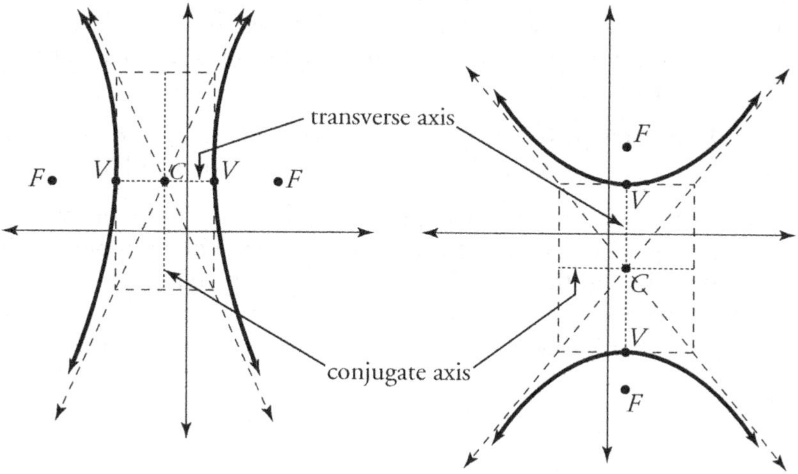


## Standard form

A hyperbola in standard form has one of the following equations:

$$\frac{(x-h)^2}{a^2} - \frac{(y-k)^2}{b^2} = 1, \quad \text{if the transverse axis is horizontal}$$

$$\frac{(y-k)^2}{a^2} - \frac{(x-h)^2}{b^2} = 1, \quad \text{if the transverse axis is vertical}$$

Each of the variables within the standard form is defined as follows:

- $(h,k)$ is the center of the hyperbola
- $a$ is the distance from the center to one of the vertices; in other words, the length of the transverse axis is $2a$
- $b$ is the distance from the center to one endpoint of the conjugate axis; in other words, the conjugate axis has length $2b$
- $c$ is the distance from the center to one of the foci, and has value $c = \sqrt{a^2 + b^2}$

Although this standard from has a lot in common with the standard form of an ellipse, there is one important distinction that is not immediately obvious. In an ellipse, you always know the major axis (with length $2a$) is the longer of the two axes. In a hyperbola, the axis whose length is $2a$ (the transverse axis) dictates which direction the hyperbola will open, but there is no guarantee that $a > b$.

**Example 8:** If a hyperbola has vertices (–3,4) and (–3,–2), and foci (–3,6) and (–3,–4), write the equation of the hyperbola in standard form.

Note that the segment connecting the foci is vertical, so use the standard form for hyperbolas with vertical transverse axes. The midpoint of the transverse axis, and therefore the center of the hyperbola, is (–3,1). The distance from the center to a focus point is $c = 5$, and the distance from the center to a vertex point is $a = 3$. Use these values to calculate $b$.

$$c = \sqrt{a^2 + b^2}$$
$$5 = \sqrt{9 + b^2}$$
$$25 = 9 + b^2$$
$$b^2 = 16$$
$$b = 4$$

Since $b$ is a distance, there is no need to consider –4 as a solution to the above equation; $a$, $b$, and $c$ will always be positive.

Fill in the values for $a$, $b$, $h$, and $k$ into standard form.

$$\frac{(y-k)^2}{a^2} - \frac{(x-h)^2}{b^2} = 1$$
$$\frac{(y-(1))^2}{3^2} - \frac{(x-(-3))^2}{4^2} = 1$$
$$\frac{(y-1)^2}{9} - \frac{(x+3)^2}{16} = 1$$

### Graphing hyperbolas

The easiest way to graph a hyperbola involves drawing the dotted rectangles pictured in Figure 8-11. Here are the steps to follow:

1. Put the hyperbola in standard form, if it is not already. Determine whether the transverse axis is horizontal or vertical.
2. Plot the center.
3. Plot the vertices, $a$ units in either direction from the center, along the transverse axis.
4. Plot the endpoints of the conjugate axis, each $b$ units away from the center. Remember, the conjugate axis is perpendicular to the transverse axis.
5. Create a rectangle such that the points you plotted in steps (3) and (4) above are the midpoints of its sides.
6. Draw the diagonals of the rectangle and extend them beyond its corners; these are the asymptotes of the hyperbola.
7. Draw branches of the hyperbola opening in the appropriate directions. They should extend out from the vertices, approaching but never touching the asymptotes.

**Example 9:** Graph the hyperbola.

$$x^2 - 2x - 25y^2 = 24$$

Put the hyperbola in standard form by completing the square for the $x$ terms.

$$x^2 - 2x + \mathbf{1} - 25y^2 = 24 + \mathbf{1}$$
$$(x - 1)^2 - 25y^2 = 25$$

Divide everything by 25, since a hyperbola in standard form is set equal to 1.

$$\frac{(x-1)^2}{25} - \frac{(y-0)^2}{1} = 1$$

This is the standard form of a hyperbola with a horizontal transverse axis, with center (1,0), $a = 5$, and $b = 1$. Use the values of $a$ and $b$ to create a rectangle whose length and width are the transverse and conjugate axes, respectively. Since the transverse axis is horizontal, the hyperbola will open left and right. See Figure 8-12.

**Figure 8-12** The graph of the hyperbola in Example 9.

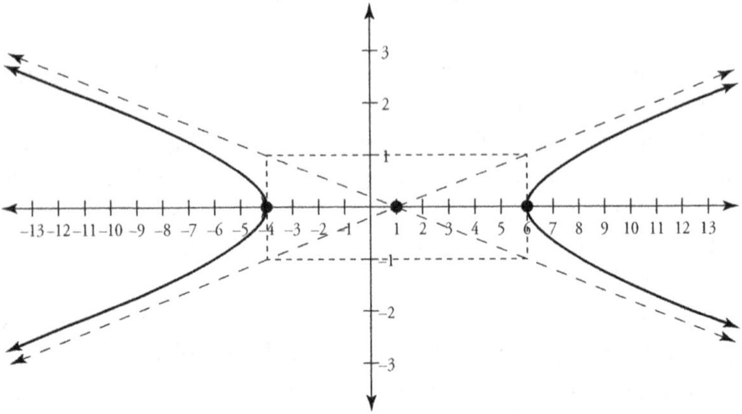


## Equations of asymptote lines

The equations for the asymptotes of a hyperbola written in standard form are:

$$y = k \pm \frac{b}{a}(x - h), \text{ if transverse axis is horizontal}$$

$$y = k \pm \frac{a}{b}(x - h), \text{ if transverse axis is vertical}$$

Note that "±" is used in the formulas because every hyperbola will possess two asymptote lines.

**Example 10:** Find the equations of the asymptotes for the hyperbola.

$$x^2 + 2x - 4y^2 + 8y - 19 = 0$$

Complete the square separately for the $x$ and $y$ terms, and put the hyperbola in standard form.

$$x^2 + 2x + \mathbf{1} - 4(y^2 - 2y + \mathbf{1}) = 19 + \mathbf{1} - \mathbf{4}$$

$$\frac{(x+1)^2}{16} - \frac{(y-1)^2}{4} = 1$$

The transverse axis is horizontal, $a = 4$, $b = 2$, and $(h,k) = (-1,1)$. Therefore, the equations of the asymptotes are

$$y = 1 + \frac{1}{2}(x + 1)$$

$$y = 1 - \frac{1}{2}(x + 1)$$

## Identifying Conic Sections

To identify a conic section, given only its equation, first transform the equation into general form:

$$Ax^2 + Bx + Cy^2 + Dy + E = 0$$

By examining the coefficients $A$ and $C$, it is a simple matter to determine what conic section the equation represents without ever having to put it into standard form.

- If $A = 0$ or $C = 0$ (meaning that either the $x^2$ or $y^2$ term is missing), the equation is a parabola.
- If $A = C$, then the equation is a circle.
- If $A \neq C$ but they have the same sign (i.e. they are both positive or both negative), then the equation is an ellipse.
- If $A \neq C$ and they have opposite signs, then the equation is a hyperbola.

For instance, in Example 10 the equation is already in general form. Since $A = 1$ and $C = -4$ (which are unequal values with opposite signs), you know that the equation is a hyperbola.

## Parametric Equations

A graph can be expressed in terms of three variables, rather than just the two variables ($x$ and $y$) you have used thus far. The third variable (usually written $t$ or $\theta$) is called the **parameter**. Every point $(x,y)$ on a normal graph can be defined in terms of a set of **parametric equations** that contain that *parameter*.

### Graphing parametric equations

Every point $(x,y)$ on a parametrically-defined graph gets its value from its defining equations

$$x = f(t) \quad y = g(t)$$

Plug a wide range of $t$-values into each equation and you'll get a corresponding coordinate pair $(f(t),g(t))$ on the graph for that particular parameter value. Remember, the parameter is not actually graphed; it is just plugged into the "$x =$" and "$y =$" equations to generate coordinate pairs.

**Example 11:** Sketch the curve generated by the parametric equations.

$$x = t + 3$$
$$y = t^2 - 2t - 1$$

Allow $t$ to take on a variety of negative and positive values, and plug each into both equations to generate points on the graph, as demonstrated in Figure 8-13.

**Figure 8-13** The graph of the parametrically-defined curve from Example 11.

| $t$ | $x = t + 3$ | $y = t^2 - 2t - 1$ | $(x,y)$ |
|---|---|---|---|
| –2 | $-2 + 3 = 1$ | $(-2)^2 - 2(-2) - 1 = 7$ | (1,7) |
| –1 | $-1 + 3 = 2$ | $(-1)^2 - 2(-1) - 1 = 2$ | (2,2) |
| 0 | $0 + 3 = 3$ | $(0)^2 - 2(0) - 1 = -1$ | (3,–1) |
| 1 | $1 + 3 = 4$ | $(1)^2 - 2(1) - 1 = -2$ | (4,–2) |
| 2 | $2 + 3 = 5$ | $(2)^2 - 2(2) - 1 = -1$ | (5,–1) |
| 3 | $3 + 3 = 6$ | $(3)^2 - 2(3) - 1 = 2$ | (6,2) |
| 4 | $4 + 3 = 7$ | $(4)^2 - 2(4) - 1 = 7$ | (4,7) |

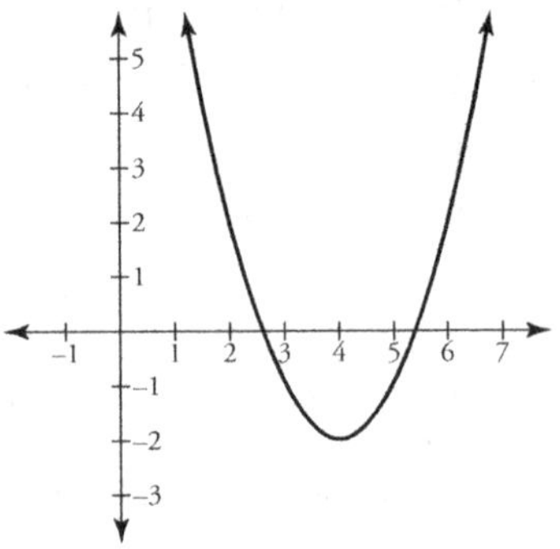


## Rewriting parametric equations

Plotting points is a tedious method for graphing parametric equations, so it's often useful to express the parametrically-defined curve in terms only of $x$ and $y$ by eliminating the parameter.

**Example 12:** Rewrite the following parametric equations by eliminating the parameter.

(a) $x = 2\cos\theta$, $y = 5\sin\theta$

In this case, the parameter is the angle $\theta$. Solve the parametric equations for $\cos\theta$ and $\sin\theta$, respectively.

$$\cos\theta = \frac{x}{2} \quad \sin\theta = \frac{y}{5}$$

Square both sides of both equations.

$$\cos^2\theta = \frac{x^2}{4} \quad \sin^2\theta = \frac{y^2}{25}$$

Consider the Pythagorean identity

$$\cos^2\theta + \sin^2\theta = 1$$

Since you have values for $\cos^2\theta$ and $\sin^2\theta$, plug them into the identity.

$$\frac{x^2}{4} + \frac{y^2}{25} = 1$$

This is the equation of an ellipse, centered at the origin, with vertical major axis of length $2 \cdot 5 = 10$ and minor axis of length $2 \cdot 2 = 4$.

(b) $x = \sqrt{t+2}, y = t+1$

Choose one of the parametric equations and solve it for $t$. It's easiest to do so with the $y$ equation.

$$y = t + 1$$
$$t = y - 1$$

Plug this $t$-value into the $x$ equation, and solve for $y$.

$$x = \sqrt{t+2}$$
$$x = \sqrt{(y-1)+2}$$
$$x = \sqrt{y+1}$$
$$x^2 = y+1$$
$$y = x^2 - 1$$

The graph of the parametric equations looks almost exactly like the parabola $y = x^2 - 1$. However, notice that it is impossible to have $x < 0$ in the original, parametric definition of the graph, since a square root must always result in a nonnegative number. Therefore, you have to adjust the transformed graph to reflect this restriction on the domain:

$$y = x^2 - 1, \text{ for } x \geq 0$$

## Polar Coordinates

Any coordinate $(x,y)$ on the normal (or **rectangular**) coordinate plane can also be expressed in terms of **polar coordinates**. Consider the point $P$ in Figure 8-14.

**Figure 8-14** A point *P* with rectangular coordinates $(x,y)$ can be expressed with polar coordinates $(r,\theta)$.

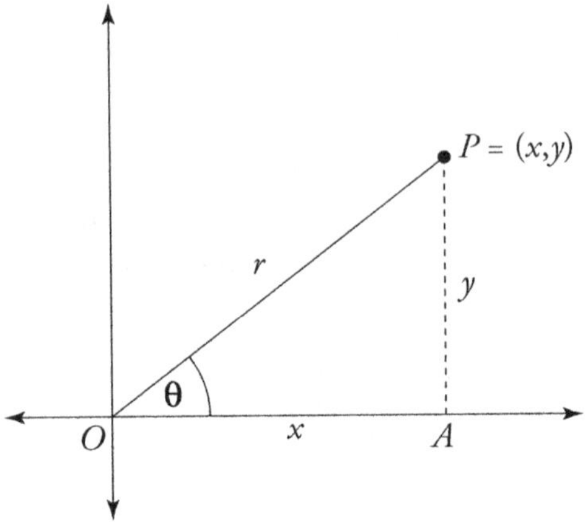


To convert $P = (x,y)$ into polar coordinates, you will need:

- The distance, $r$, from a fixed point $O$ (called the **pole**) to $P$. For the sake of simplicity, the origin is usually chosen as the *pole*, and in fact you should assume that to be true unless told otherwise.
- The angle, $\theta$, with initial side $\overrightarrow{OA}$ (called the **polar axis**) and terminal side $\overrightarrow{OP}$, such that $\theta > 0$ implies counterclockwise rotation and $\theta < 0$ implies clockwise rotation. Traditionally, the positive $x$-axis is chosen to be the *polar axis*.

These values, written as the coordinate pair $(r,\theta)$ represent the rectangular point $P$ in polar form. Note that if $r < 0$, you should travel in the direction opposite the terminal side of $\theta$.

**Example 13:** Graph the polar coordinates $A = (2,\frac{3\pi}{4})$, $B = (-3,\pi)$, and $C = (-4,-\frac{5\pi}{6})$.

Unlike rectangular coordinates, polar coordinates are not unique, thanks to negative and coterminal angles. For instance, consider point $C$ in Figure 8-15. It could also be expressed as $(4,\frac{\pi}{6})$, $(4,\frac{13\pi}{6})$ or $(-4,-\frac{17\pi}{6})$.

**Figure 8-15** The graphs of polar coordinates *A*, *B*, and *C* from Example 13.

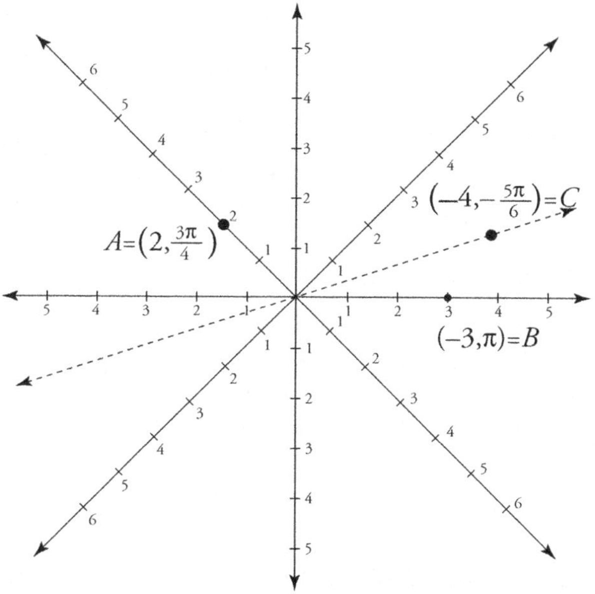


## Converting between polar and rectangular coordinates

Because triangle *AOP* in Figure 8-14 is a right triangle, you can draw the following trigonometric conclusions:

- $\cos\theta = \frac{x}{r}$
- $\sin\theta = \frac{y}{r}$
- $\tan\theta = \frac{y}{x}$
- $x^2 + y^2 = r^2$

The first two of those four conclusions allow you to convert from polar to rectangular coordinates. Solve them for *x* and *y*, respectively, to get the conversion equations:

$$x = r\cos\theta \quad y = r\sin\theta$$

The final two of the above bullets will assist you when converting the other direction, from rectangular to polar coordinates.

**Example 14:** Convert to rectangular coordinates: $(-2,\frac{3\pi}{2})$.

Apply the conversion equations.

$$x = r\cos\theta \qquad y = r\sin\theta$$

$$x = (-2)\cos\frac{3\pi}{2} \qquad y = (-2)\sin\frac{3\pi}{2}$$

$$x = (-2)(0) \qquad y = (-2)(-1)$$

$$x = 0 \qquad y = 2$$

The rectangular coordinates are (0,2).

**Example 15:** Convert to polar coordinates: (1,–2).

Begin by applying the Pythagorean polar coordinate conclusion.

$$r^2 = x^2 + y^2$$

$$r^2 = 1^2 + (-2)^2 = 5$$

$$r = \sqrt{5}$$

Now apply the tangent formula to find θ.

$$\tan\theta = \frac{y}{x}$$

$$\tan\theta = -\frac{2}{1}$$

$$\theta = \arctan(-2)$$

There is no obvious unit circle angle whose tangent value equals –2, so either leave θ as arctan (–2) or use a calculator to evaluate its value. Therefore, one possible polar coordinate representation of the point (1,–2) is

$$\left(\sqrt{5}, \arctan(-2)\right)$$

## Converting between polar and rectangular equations

To convert a rectangular equation to polar form, use the same conversion formulas from Example 14 to replace the values of $x$ and $y$.

**Example 16:** Rewrite in polar form: $x - 2y = 4$.

Replace $x$ with $r\cos\theta$ and $y$ with $r\sin\theta$.

$$r\cos\theta - 2(r\sin\theta) = 4$$

Polar equations are usually written in terms of θ, so solve for $r$.

$$r(\cos\theta - 2\sin\theta) = 4$$

$$r = \frac{4}{\cos\theta - 2\sin\theta}$$

To convert a polar equation into rectangular form, the technique is not as straightforward; your method will depend upon the problem. You may need any or all of the four trigonometric conclusions generated by Figure 8-14.

**Example 17:** Rewrite in rectangular form: $r = 2\sin\theta$.

Multiply both sides by $r$.

$$r^2 = 2r\sin\theta$$

Remember that $r^2 = x^2 + y^2$ and $r\sin\theta = y$; make those substitutions.

$$x^2 + y^2 = 2y$$

This is the equation of a circle. Put it in standard form.

$$x^2 + y^2 - 2y = 0$$

$$x^2 + (y^2 - 2y + 1) = 1$$

$$x^2 + (y - 1)^2 = 1$$

The circle's center is (0,1) and it has radius 1.

If you are unable to easily convert a polar equation into rectangular form but are still required to graph it, plug a wide range of θ's from the unit circle into the equation and plot the corresponding points. (Just like you can pick a wide range of $x$'s and plug them into a rectangular equation to get the corresponding $y$'s and therefore the coordinates of points on a rectangular graph.)

## Chapter Checkout

### Q&A

1. Identify the following conic sections and put them in standard form.
   (a) $36x^2 - y^2 - 8y + 20 = 0$
   (b) $x^2 + x + y^2 - 6y + \frac{1}{4} = 0$
   (c) $3x^2 - 18x - 4y + 29 = 0$
   (d) $2x^2 + y^2 = 14$

**2.** Find the coordinates of the foci:

$$9x^2 - 36x + 4y^2 - 56y + 196 = 0$$

**3.** Rewrite the parametric equations by eliminating the parameter:

$$x = e^t \quad y = t^3 - 2$$

**4.** Rewrite the polar coordinates $\left(-2, \frac{5\pi}{4}\right)$ in rectangular form.

**Answers: 1. (a)** hyperbola, $\frac{(y+4)^2}{36} - \frac{x^2}{1} = 1$
**(b)** circle, $\left(x + \frac{1}{2}\right)^2 + (y-3)^2 = 9$ **(c)** parabola, $y = \frac{3}{4}(x-3)^2 + \frac{1}{2}$
**(d)** ellipse, $\frac{x^2}{7} + \frac{y^2}{14} = 1$ **2.** $\left(2, 7+\sqrt{5}\right), \left(2, 7-\sqrt{5}\right)$ **3.** $y = (\ln x)^3 - 2$
**4.** $\left(\sqrt{2}, \sqrt{2}\right)$

# Chapter 9

# MATRICES AND SYSTEMS OF EQUATIONS

## Chapter Check-In

- ❑ Determining solutions to systems of equations
- ❑ Performing operations on matrices
- ❑ Solving systems of equations using augmented matrices
- ❑ Writing matrices in row-echelon and reduced row-echelon form
- ❑ Graphing the solution to systems of inequalities

In this chapter, your focus shifts from working with a single function, inequality, equation, or graph to working with multiple equations at once. Your primary goal will be to determine solutions (in the form of coordinates or regions) that make two or more equations or inequalities true.

## Systems of Equations

A **system of equations** is a set of equations for which you are seeking a single set of coordinates that makes all of the equations in the set true. Consider the system of equations

$$\begin{cases} 3x - y = 7 \\ x + 5y = -19 \end{cases}$$

The solution to this system is the $(x,y)$ coordinate pair $(1,-4)$, because that is the only $x$- and $y$-value which, together, make both equations true.

$$3(1) - (-4) = 7 \qquad 1 + 5(-4) = -19$$
$$3 + 4 = 7 \qquad 1 - 20 = -19$$

A system that has no solutions is described as **inconsistent**, whereas a system with an infinite number of solutions is said to be **dependent**.

Graphically speaking, the solution to a system of equations is the point or points at which the equations in the system intersect. In Figure 9-1, note that the lines $3x - y = 7$ and $x + 5y = -19$ intersect at the coordinate pair representing the solution: (1,–4).

**Figure 9-1** The solution to the system of equations is the intersection point of its two graphs.

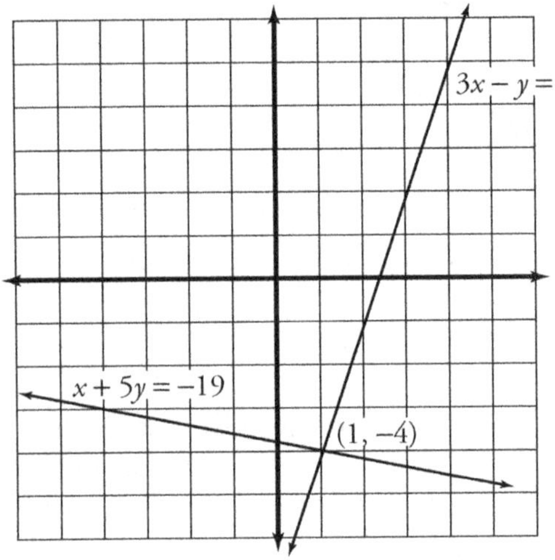


Without the use of computer or calculator graphing technology, solving systems of equations by determining where their graphs intersect is not very accurate. Graphs drawn by hand, even on graphing paper, do not have the built-in precision to allow you to trust any conclusions draw from them (especially if the coordinates of the solutions are not integers).

## Two-Variable Linear Systems

There are two methods most often used to solve systems of linear equations containing only two variables. Although both work for any system of equations, some problems are more suited for one than the other, meaning that you may reach a solution faster and with less computation if you choose between the two carefully.

## Substitution method

If one of the equations can easily be solved for a variable, the system is a good candidate for the substitution method. Here are the steps to follow:

1. Solve an equation for one of its variables.
2. Substitute that variable's value into the other equation and find the value of the remaining variable.
3. Plug the resulting value into the equation you originally manipulated to complete the ordered pair.

**Example 1:** Solve the following systems of equations.

(a) $\begin{cases} 2x - y = 4 \\ 8x + 3y = -5 \end{cases}$

Since the $y$-term in the first equation has a coefficient of $-1$, it's easy to solve for it; the result should have no fractions in it and therefore be easier to manipulate.

$$y = 2x - 4$$

Substitute for $y$ in the other equation.

$$8x + 3(2x - 4) = -5$$
$$8x + 6x - 12 = -5$$
$$14x = 7$$
$$x = \frac{1}{2}$$

Now that you know the $x$-value of the solution, plug it into the equation you previously solved for $y$.

$$y = 2\left(\frac{1}{2}\right) - 4$$
$$y = 1 - 4 = -3$$

The solution to this system is $\left(\frac{1}{2}, -3\right)$.

(b) $\begin{cases} x - 2y = 3 \\ 4x - 12 = 8y \end{cases}$

Solve for first equation for $x$ and plug it into the second equation.

$$x = 2y + 3$$
$$4(2y + 3) - 12 = 8y$$
$$8y + 12 - 12 = 8y$$
$$8y = 8y$$

This equation is true no matter what value you plug in for $y$, so this is a dependent system, and it has an infinite number of solutions.

## Elimination method

Remember that any equation multiplied by (or divided by) a nonzero real number will still keep its original solution(s) although the equation may look different. This is the foundational principle of the elimination method, whose steps are as follows:

1. Choose a real number (or numbers) such that if you multiply one (or both) of the equations in the system by the number(s), and then add the equations, one of the variables in the system is eliminated.
2. Solve the result of the equation sum that results from Step (1).
3. Plug the value you find into either of the two original equations to complete the ordered pair.

There is an alternative to choosing a real number to eliminate a variable. First, put the equations in standard form (such that $A > 0$ and $D > 0$).

$$\begin{cases} Ax + By = C \\ Dx + Ey = F \end{cases}$$

Multiply the first equation by $D$ and the second by $-A$, then add the equations together (by combining like terms). This will eliminate the $x$ variable, no matter what the coefficients are.

**Example 2:** Solve the systems of equations.

(a) $\begin{cases} 2x - 4y = \dfrac{51}{2} \\ 8x + 3y = -12 \end{cases}$

Multiply all the terms in the first equation by (–4).

$$-4(2x - 4y = \frac{51}{2})$$
$$-8x + 16y = -102$$

Add the new form of the first equation to the second equation, noting that the *x*-terms will cancel out. (That's why the multiple of –4 was chosen for the first equation.)

$$\begin{array}{rrr} -8x & +16y & =-102 \\ 8x & +3y & =-12 \\ \hline & 19y & =-114 \\ & y & =-6 \end{array}$$

Substitute this value for *y* in either of the two original equations.

$$8x + 3(-6) = -12$$

$$8x - 18 = -12$$

$$8x = 6$$

$$x = \frac{3}{4}$$

The solution to this system is $\left(\frac{3}{4}, -6\right)$.

(b) $\begin{cases} x - 2y = -4 \\ -3x + 6y = 1 \end{cases}$

Multiply the top equation by 3 and add the equations together.

$$\begin{array}{rrr} 3x & -6y & =-12 \\ -3x & +6y & = 1 \\ \hline & 0 & =-11 \end{array}$$

This is a false statement, and no $(x,y)$ pair can make it true. Therefore, this system is inconsistent, as it has no solutions.

## Nonlinear Systems of Equations

The substitution and elimination methods still apply to systems of equations even if the equations aren't linear. Whereas substitution works more often than elimination for nonlinear systems, when elimination works, it works quickly and effectively.

**Example 3:** Solve the systems of equations.

(a) $\begin{cases} x^2 - 4x + 2y + 6 = 0 \\ 2x - y = 11 \end{cases}$

Use the substitution method; solve the second equation for *y*.

$$y = 2x - 11$$

Substitute this value for $y$ into the first equation and solve.

$$x^2 - 4x + 2(2x - 11) + 6 = 0$$

$$x^2 - 4x + 4x - 22 + 6 = 0$$

$$x^2 = 16$$

$$x = \pm 4$$

Plug both values of $x$ into the equation you solved for $y$.

$$(x = -4): y = 2(-4) - 11 = -19$$

$$(x = 4): y = 2(4) - 11 = -3$$

There are two coordinate pair which make up the solution: (–4,–19) and (4,–3).

(b) $\begin{cases} 4x^2 + 9y^2 = 36 \\ x^2 + y^2 = 9 \end{cases}$

Perform the elimination method; multiply the second equation by –4 and add the two equations together.

$$\begin{array}{rrcr} 4x^2 & +9y^2 & = & 36 \\ -4x^2 & -4y^2 & = & -36 \\ \hline & 5y^2 & = & 0 \\ & y & = & 0 \end{array}$$

Substitute $y = 0$ into the second equation in its original form.

$$x^2 + 0^2 = 9$$

$$x = \pm 3$$

The solution to this system consists of the coordinates (3,0) and (–3,0).

## Characteristics of Matrices

A **matrix** is a rectangular collection of numbers, arranged in rows and columns, surrounded by a single set of brackets on either side. The **order** of a *matrix* describes how many rows and columns are contained within. The *matrix* below has order 3 × 4 (read "3 by 4"), because it has 3 rows and 4 columns.

$$\begin{bmatrix} a_{11} & a_{12} & a_{13} & a_{14} \\ a_{21} & a_{22} & a_{23} & a_{24} \\ a_{31} & a_{32} & a_{33} & a_{34} \end{bmatrix}$$

Each number within the matrix is called an **element** (or an **entry**), and it is designated by two subscripts, $a_{ij}$, where $i$ is the number representing the element's row and $j$ represents its column. Therefore, $a_{21}$ is the element in the second row (from the top) and the first column (from the left).

An italicized capital letter is usually used to designate a matrix; you may include the order of the matrix as a subscript if you wish, for the sake of clarity: $A_{3\times4}$. You can also represent a matrix by writing one of its generic elements in double subscript notation and surrounding it in either brackets or parentheses: $[a_{ij}]$.

If the matrix has an equal number of rows and columns, it is said to be a **square matrix**. A *square matrix* of order $n \times n$ has a **diagonal** containing the elements $a_{11}, a_{22}, a_{33}, \ldots, a_{nn}$.

If a matrix consists of only one row, it is called a **row matrix**; similarly, a matrix consisting of only one column is called a **column matrix**. A matrix of any order which contains only zeros as elements is called a **zero matrix**.

## Basic Matrix Operations

You should be able to add, subtract, and multiply matrices. Furthermore, you should also be able to multiply a matrix by a scalar (numeric) value.

### Adding matrices

The sum of two matrices, $A = [a_{ij}]$ and $B = [b_{ij}]$, that have the same order is $C = [c_{ij}]$, where $c_{ij} = a_{ij} + b_{ij}$. In other words, each element in $C$ is equal to the sum of the elements in $A$ and $B$ that are located in corresponding spots. Therefore, you know that $c_{25}$ (the element in the second row and fifth column of $C$) is equal to the sum of $a_{25} + b_{25}$. You can only add matrices if they have the same order.

**Example 4:** Find the sum of the matrices.

$$\begin{bmatrix} 2 & -4 \\ -5 & 7 \end{bmatrix} + \begin{bmatrix} 0 & 6 \\ 1 & -3 \end{bmatrix}$$

Add the elements whose locations correspond.

$$\begin{bmatrix} 2+0 & -4+6 \\ -5+1 & 7+(-3) \end{bmatrix}$$

$$\begin{bmatrix} 2 & 2 \\ -4 & 4 \end{bmatrix}$$

Note that matrix addition is both commutative and associative.

## Scalar multiplication

If $n$ is a scalar (a real number) and $A = [a_{ij}]$, then

$$n \cdot A = [n \cdot a_{ij}]$$

In other words, if a matrix if multiplied by a scalar, then multiply every one of its elements by that scalar.

**Example 5:** Given the following matrices, evaluate $2A + B$.

$$A = \begin{bmatrix} 2 & 1 & 4 & -3 \\ -5 & -2 & 0 & 6 \end{bmatrix} \qquad B = \begin{bmatrix} -1 & -6 & 5 & -10 \\ 8 & 2 & 4 & -3 \end{bmatrix}$$

Begin by multiplying every element in $A$ by the scalar 2.

$$2A = \begin{bmatrix} 4 & 2 & 8 & -6 \\ -10 & -4 & 0 & 12 \end{bmatrix}$$

Now add $2A$ and $B$ together.

$$2A + B = \begin{bmatrix} 4+(-1) & 2+(-6) & 8+5 & -6+(-10) \\ -10+8 & -4+2 & 0+4 & 12+(-3) \end{bmatrix}$$

$$= \begin{bmatrix} 3 & -4 & 13 & -16 \\ -2 & -2 & 4 & 9 \end{bmatrix}$$

## Subtracting matrices

In order to perform matrix subtraction, multiply the second matrix by a scalar value of (–1) and add the matrices:

$$A - B = A + (-B)$$

Since subtraction really boils down to addition, both matrices must have the same order in order to subtract them.

**Example 6:** Given the following matrices, evaluate $A - 2B$.

$$A = \begin{bmatrix} 4 & -3 \\ 2 & -1 \end{bmatrix} \quad B = \begin{bmatrix} 5 & 7 \\ -9 & 6 \end{bmatrix}$$

This subtraction problem is equivalent to the addition (and scalar multiplication) problem of $A + (-2B)$. Begin by calculating $-2B$.

$$-2B = \begin{bmatrix} -10 & -14 \\ 18 & -12 \end{bmatrix}$$

Now add $A + (-2B)$.

$$A + (-2B) = \begin{bmatrix} -6 & -17 \\ 20 & -13 \end{bmatrix}$$

## Multiplying matrices

Two matrices need not have the same order if you want to find their product. However, in order for the product of two matrices, $A \cdot B$ to exist, the number of *columns* in $A$ must equal the number of *rows* in $B$. The pattern for calculating matrix products is a bit more complex than scalar multiplication and matrix addition.

If $A_{m \times n} = [a_{ij}]$ and $B_{n \times p} = [b_{ij}]$, then $C = A \cdot B = [c_{ij}]$ is an $m \times p$ matrix defined as follows.

1. Choose an unknown value $c_{ij}$ in $C$ you'd like to evaluate.
2. Multiply each element in the *i*th row of $A$ (from left to right) times its corresponding element in the *j*th column of $B$ (from top to bottom). In other words, $a_{i1}$ is multiplied by $b_{1j}$, $a_{i2}$ is multiplied by $b_{2j}$, and so forth until you simultaneously reach the end of the row and column. (This is why the number of columns in $A$ must match the number of rows in $B$.)
2. The sum of all those products is equal to $c_{ij}$.
3. Repeat the process for all elements of $C$.

Consider the matrices $A$ and $B$.

$$A = \begin{bmatrix} a_{11} & a_{12} \\ a_{21} & a_{22} \end{bmatrix} \quad B = \begin{bmatrix} b_{11} & b_{12} & b_{13} \\ b_{21} & b_{22} & b_{23} \end{bmatrix}$$

Since $A$ has 2 columns and $B$ has 2 rows, the product of these matrices exists, and will be $2 \times 3$.

$$A \cdot B = \begin{bmatrix} a_{11} \cdot b_{11} + a_{12} \cdot b_{21} & a_{11} \cdot b_{12} + a_{12} \cdot b_{22} & a_{11} \cdot b_{13} + a_{12} \cdot b_{23} \\ a_{21} \cdot b_{11} + a_{22} \cdot b_{21} & a_{21} \cdot b_{12} + a_{22} \cdot b_{22} & a_{21} \cdot b_{13} + a_{22} \cdot b_{23} \end{bmatrix}$$

Note that matrix multiplication is associative, is not commutative (because of the row/column restriction), but is distributive over matrix addition, so $A(B + C) = AB + AC$.

**Example 7:** Calculate the products of the matrices.

(a) $\begin{bmatrix} 4 & 0 \\ -2 & 7 \\ 3 & 1 \end{bmatrix} \cdot \begin{bmatrix} 6 & -1 \\ -5 & 2 \end{bmatrix}$

The result will be a $3 \times 2$ matrix.

$$\begin{bmatrix} 4 \cdot 6 + 0(-5) & 4(-1) + 0 \cdot 2 \\ -2 \cdot 6 + 7(-5) & (-2)(-1) + 7 \cdot 2 \\ 3 \cdot 6 + 1(-5) & 3(-1) + 1 \cdot 2 \end{bmatrix} = \begin{bmatrix} 24 & -4 \\ -47 & 16 \\ 13 & -1 \end{bmatrix}$$

(b) $\begin{bmatrix} 1 & 3 & -5 \\ -4 & 7 & 6 \end{bmatrix} \cdot \begin{bmatrix} 2 & -8 \\ 9 & -1 \\ 0 & 4 \end{bmatrix}$

The result will be a $2 \times 2$ matrix.

$$\begin{bmatrix} 1 \cdot 2 + 3 \cdot 9 + (-5)(0) & 1(-8) + 3(-1) + (-5)(4) \\ (-4)(2) + 7 \cdot 9 + 6 \cdot 0 & (-4)(-8) + 7(-1) + 6(4) \end{bmatrix}$$

$$= \begin{bmatrix} 29 & -31 \\ 55 & 49 \end{bmatrix}$$

## Solving Systems of Equations with Matrices

If a system of equations contains 3 or more different variables, it becomes difficult to solve via elimination or substitution. One very orderly method for finding solutions to such systems is to create matrices based on their coefficients, called **coefficient matrices.**

Consider the system

$$\begin{cases} x - 5y + 3z = -10 \\ 2x + y - z = 8 \\ -x + 3y + 7y = -30 \end{cases}$$

You can find the solution $(x, y, z)$ by manipulating the **augmented matrix**.

$$\left[\begin{array}{ccc|c} 1 & -5 & 3 & -10 \\ 2 & 1 & -1 & 8 \\ -1 & 3 & 7 & -30 \end{array}\right]$$

This matrix is said to be *augmented* because it is more than just a coefficient matrix; it also includes an extra column (on the far right) that contains the constants on the right side of the equal signs in the system. The column(s) which make this an augmented matrix are usually separated from the coefficient matrix by a thin or dotted line, as shown above.

## Gaussian and Gauss-Jordan elimination

Your goal, when manipulating matrices to solve systems of equations, is to get elements of 1 in the diagonal ($a_{11}$, $a_{22}$, $a_{33}$, ...) of the coefficient matrix. In addition, you want all of the elements left of the diagonal to be 0. If there are any rows containing only 0's, they should be placed at the very bottom of the matrix.

$$\left[\begin{array}{cccc|c} 1 & a & b & c & d \\ 0 & 1 & e & f & g \\ 0 & 0 & 1 & h & j \\ 0 & 0 & 0 & 0 & 0 \end{array}\right]$$

This process is called **Gaussian elimination** and the result is a matrix in **row-echelon form**. When matrices are in *row-echelon form*, it is a simple matter of back-substitution to reach a solution.

You can take this process one step farther and force all of the non-diagonal elements in the coefficient matrix to be 0's as well. This process is called **Gauss-Jordan elimination**, and the result is a matrix in **reduced row-echelon form**, like the matrix below. (Although that matrix has no rows that contain only zeros, such rows should again be placed at the bottom of the matrix.)

$$\left[\begin{array}{cccc|c} 1 & 0 & 0 & 0 & a \\ 0 & 1 & 0 & 0 & b \\ 0 & 0 & 1 & 0 & c \\ 0 & 0 & 0 & 1 & d \end{array}\right]$$

The solutions for this matrix stand out visibly in the rightmost column of constants. Though the process of *Gauss-Jordan elimination* is a bit more tedious than mere Gaussian elimination, no back-substitution is necessary to reach a solution.

Note that a matrix in *row-echelon* or *reduced row-echelon form* need not contain row(s) of 0's, but if it does, any such row should appear at the bottom of the matrix. Also, matrices that have a greater number of columns than rows can still be put in either form; they do not have to be square. The 1 elements still appear in the same place ($a_{11}$, $a_{22}$, $a_{33}$, etc.), although this set is not technically called the diagonal if the matrix does not have a matching number of rows and columns.

## Matrix row operations

In order to reach row-echelon or reduced row-echelon form, you are allowed to manipulate the rows in a matrix as follows:

1. Multiply a row by a constant (except 0).
2. Switch the positions of two rows.
3. Replace a row by its sum with another row.

You can also combine these row operations, so you can multiply a row by a constant and add it to another row. In the examples that follow, this will be indicated by arrow notation. For example, "$-2R_2 + R_3 \rightarrow R_3$" means "Multiply the second row by –2, add the result to the third row, and write the result in the third row, replacing its previous values."

**Example 8:** Solve the system by rewriting it as an augmented matrix in row-echelon form.

$$\begin{cases} x-5y+3z=-10 \\ 2x+y-z=8 \\ -x+3y+7y=-30 \end{cases}$$

As indicated earlier in this section, this system can be rewritten as this augmented matrix:

$$\left[\begin{array}{rrr|r} 1 & -5 & 3 & -10 \\ 2 & 1 & -1 & 8 \\ -1 & 3 & 7 & -30 \end{array}\right]$$

Perform these operations: $-2R_1 + R_2 \rightarrow R_2$ and $R_1 + R_3 \rightarrow R_3$. The net result is a 1 at the top of the left column with 0's beneath it.

$$\left[\begin{array}{ccc|c} 1 & -5 & 3 & -10 \\ 0 & 11 & -7 & 28 \\ 0 & -2 & 10 & -40 \end{array}\right]$$

Interchange $R_2$ and $R_3$, then divide row 2 by –2. This changes its second-column element to 1.

$$\left[\begin{array}{ccc|c} 1 & -5 & 3 & -10 \\ 0 & 1 & -5 & 20 \\ 0 & 11 & -7 & 28 \end{array}\right]$$

To eliminate the 11 in position $a_{32}$, perform $-11R_2 + R_3 \rightarrow R_3$.

$$\left[\begin{array}{ccc|c} 1 & -5 & 3 & -10 \\ 0 & 1 & -5 & 20 \\ 0 & 0 & 48 & -192 \end{array}\right]$$

Divide $R_3$ by 48 to get $a_{33} = 1$.

$$\left[\begin{array}{ccc|c} 1 & -5 & 3 & -10 \\ 0 & 1 & -5 & 20 \\ 0 & 0 & 1 & -4 \end{array}\right]$$

This matrix can now be converted back into a system of equations. Remember, columns 1, 2, and 3 correspond to the coefficients of $x$, $y$, and $z$ respectively.

$$\begin{cases} x - 5y + 3z = -10 \\ y - 5z = 20 \\ z = -4 \end{cases}$$

Since you now know that $z = -4$, plug that into the second equation to find $y$.

$$y - 5(-4) = 20$$
$$y + 20 = 20$$
$$y = 0$$

Now plug $y = 0$ and $z = -4$ into the first equation to calculate $x$.

$$x - 5(0) + 3(-4) = -10$$
$$x - 12 = -10$$
$$x = 2$$

The solution is $(x,y,z) = (2,0,-4)$.

**Example 9:** Solve the system by rewriting it as an augmented matrix in reduced row-echelon form.

$$\begin{cases} 2x + 3y = 16 \\ 4x - y = -3 \end{cases}$$

Rewrite as a matrix and divide $R_1$ by 2 to begin the diagonal of 1's.

$$\left[\begin{array}{cc:c} 1 & 3/2 & 8 \\ 4 & -1 & -3 \end{array}\right]$$

Eliminate the 4 in position $a_{21}$ with the operation $-4R_1 + R_2 \rightarrow R_2$.

$$\left[\begin{array}{cc:c} 1 & 3/2 & 8 \\ 0 & -7 & -35 \end{array}\right]$$

Divide $R_2$ by $-7$ to achieve row-echelon form.

$$\left[\begin{array}{cc:c} 1 & 3/2 & 8 \\ 0 & 1 & 5 \end{array}\right]$$

To achieve *reduced* row-echelon form, element $a_{12}$ must be a 0, so that all non-diagonal entries in the coefficient matrix are 0. The operation $-\frac{3}{2}R_2 + R_1 \rightarrow R_1$ will achieve that goal.

$$\left[\begin{array}{cc:c} 1 & 0 & 1/2 \\ 0 & 1 & 5 \end{array}\right]$$

If you rewrite this matrix as a system of equations, you get the solution with no further substitution or simplification.

$$x = \frac{1}{2}, y = 5$$

### Systems with infinitely many solutions

Although both Examples 8 and 9 worked out nicely, both providing complete solutions, it is possible that the system won't have a single solution but an infinite number of answers. That's not caused by the use of matrices, but augmented matrices do make such an occurrence easier to spot. Systems with infinite solutions will not transform into a matrix with a valid diagonal of 1's, whether it's because the coefficient matrix isn't square or one of the rows cancels out and becomes a row of 0's (in which it is impossible to introduce a 1 to complete the diagonal).

For example, if you try to solve the system

$$\begin{cases} 3x + 9y - 15z = 6 \\ 2x - 10z = 4 \end{cases}$$

using Gauss-Jordan elimination, you'll get the matrix

$$\left[\begin{array}{ccc|c} 1 & 3 & -5 & 2 \\ 0 & 1 & 0 & 0 \end{array}\right]$$

Since there is no third row, the coefficient matrix is not square. You cannot find a specific $z$, and there will be infinitely many solutions. To write the set of solutions, set $z = c$, where $c$ is any real number, and back-substitute like you did in Example 8.

$$z = c$$
$$y = 0$$
$$x + 3(0) - 5c = 2$$
$$x = 2 + 5c$$

The solution is $(2 + 5c, 0, c)$, where $c$ is a real number. To convince yourself that the solution is valid, pick a $c$ value and plug it into the system to check. For instance, consider $c = 5$. Plugging this into the equations of the system results in true statements:

$$c = 5 \text{ corresponds to solution } (2 + 5(5), 0, 5) = (27, 0, 5)$$
$$3(27) + 9(0) - 15(5) = 81 - 75 = 6 \checkmark$$
$$2(27) - 10(5) = 54 - 50 = 4 \checkmark$$

## Inverse Matrices

In Chapter 1, you learned that addition and multiplication of real numbers have inverse properties which (when applied) cancel out a value, effectively returning that operation's identity element:

$$a + (-a) = 0 \quad a \cdot \frac{1}{a} = 1$$

So, too, a square matrix $A_{m \times m}$ may have an **inverse matrix** $A^{-1}{}_{m \times m}$ such that the product $A \cdot A^{-1}$ equals the $m \times m$ **identity matrix**.

You have already (albeit unknowingly) dealt with the *identity matrix* when you were applying Gauss-Jordan elimination. It is the square matrix which contains all 0 elements except for its diagonal, which contains only 1 elements. For instance, the $4 \times 4$ identity matrix is

$$\begin{bmatrix} 1 & 0 & 0 & 0 \\ 0 & 1 & 0 & 0 \\ 0 & 0 & 1 & 0 \\ 0 & 0 & 0 & 1 \end{bmatrix}$$

The identity matrix acts just like any other identity element; if any square matrix $A$ is multiplied by the identity matrix of the same order, the result is $A$.

### Calculating inverse matrices

The inverse matrix $A^{-1}$ is unique for every matrix $A$; that is, if a matrix has an inverse, it has only one. Note that only square matrices can possess inverses.

1. Create an augmented matrix $A_0 = [A \,\vdots\, I]$, where $I$ is the identity matrix with the same order as $A$.
2. Manipulate $A_0$ so that the left-hand square matrix ($A$) is in reduced row-echelon form. If you cannot do so (perhaps one of the rows becomes all 0's), then $A$ is **singular**, meaning it has no inverse.
3. The right-hand square matrix (formerly the identity matrix, before you put $A$ into reduced row-echelon form) is $A^{-1}$.

**Example 10:** Find the inverses of the matrices.

(a) $M = \begin{bmatrix} 1 & -2 \\ -3 & 5 \end{bmatrix}$

Create an augmented matrix $M_0$ containing the $2 \times 2$ identity matrix on its right side and $M$ on its left.

$$M_0 = \left[\begin{array}{rr|rr} 1 & -2 & 1 & 0 \\ -3 & 5 & 0 & 1 \end{array}\right]$$

Begin with $3R_1 + R_2 \rightarrow R_2$ to try and reach reduced row-echelon form.

$$M_0 = \left[\begin{array}{cc:cc} 1 & -2 & 1 & 0 \\ 0 & -1 & 3 & 1 \end{array}\right]$$

The operations $-2R_2 + R_1 \rightarrow R_1$ followed by $-R_2 \rightarrow R_2$ will turn the left-hand matrix into an identity matrix, which is your goal in reduced row-echelon form.

$$M_0 = \left[\begin{array}{cc:cc} 1 & 0 & -5 & -2 \\ 0 & 1 & -3 & -1 \end{array}\right]$$

The right-hand matrix is the inverse matrix of $M$.

$$M^{-1} = \begin{bmatrix} -5 & -2 \\ -3 & -1 \end{bmatrix}$$

(b) $A = \begin{bmatrix} 2 & 1 & 0 \\ -1 & 0 & 2 \\ -2 & -1 & 1 \end{bmatrix}$

Begin by augmenting this matrix with a $3 \times 3$ identity matrix.

$$A_0 = \left[\begin{array}{ccc:ccc} 2 & 1 & 0 & 1 & 0 & 0 \\ -1 & 0 & 2 & 0 & 1 & 0 \\ -2 & -1 & 1 & 0 & 0 & 1 \end{array}\right]$$

Put the left-hand square matrix in reduced row-echelon form; you will wind up with

$$A_0 = \left[\begin{array}{ccc:ccc} 1 & 0 & 0 & 2 & -1 & 2 \\ 0 & 1 & 0 & -3 & 2 & -4 \\ 0 & 0 & 1 & 1 & 0 & 1 \end{array}\right]$$

$A^{-1}$ is the $3 \times 3$ matrix on the right-hand side of the dashed line.

### Solving matrix equations

You've already seen Gaussian and Gauss-Jordan elimination used to solve a system of equations, but you have additional options available to you for solving systems, including matrix equations.

Consider the equation $4x = 8$. To solve this for $x$, you must cancel the 4 by means of its multiplicative inverse, $\frac{1}{4}$:

$$\left(4 \cdot \frac{1}{4}\right)x = \left(\frac{1}{4}\right)8$$
$$x = 2$$

Similarly, you can solve the matrix equation

$$AX = B,$$

where $A$ is a coefficient matrix, $X$ is a column matrix consisting of the corresponding variables, and $B$ is a column matrix containing all the constants. To solve the matrix equation for $X$, multiply each side of the equal sign by $A^{-1}$:

$$(A^{-1} \cdot A)X = (A^{-1})B$$
$$X = A^{-1} \cdot B$$

**Example 11:** Solve the system by first translating it into a matrix equation and then solving that equation.

$$\begin{cases} 2x + y - 3z = 2 \\ x + 4z = 7 \\ -4x - 2y + z = -9 \end{cases}$$

This system is equivalent to the following matrix equation:

$$\begin{bmatrix} 2 & 1 & -3 \\ 1 & 0 & 4 \\ -4 & -2 & 1 \end{bmatrix}\begin{bmatrix} x \\ y \\ z \end{bmatrix} = \begin{bmatrix} 2 \\ 7 \\ -9 \end{bmatrix}$$

Calculate the inverse of the $3 \times 3$ matrix and multiply it on both sides of the equation to isolate the variable matrix.

$$\begin{bmatrix} x \\ y \\ z \end{bmatrix} = \begin{bmatrix} 8/5 & 1 & 4/5 \\ -17/5 & -2 & -11/5 \\ -2/5 & 0 & -1/5 \end{bmatrix}\begin{bmatrix} 2 \\ 7 \\ -9 \end{bmatrix}$$

$$\begin{bmatrix} x \\ y \\ z \end{bmatrix} = \begin{bmatrix} 16/5 + 35/5 + 36/5 \\ -34/5 - 70/5 + 99/5 \\ -4/5 + 0 + 9/5 \end{bmatrix} = \begin{bmatrix} 3 \\ -1 \\ 1 \end{bmatrix}$$

The solution to the system is $x = 3, y = -1, z = 1$.

## Determinants

A **determinant** is a real number that is defined for any square matrix $A$; it is expressed either as det ($A$) or $|A|$. (You can also indicate a *determinant* by writing the matrix itself bounded by single bars, instead of brackets.)

If the square matrix $A$ is $1 \times 1$ (the matrix has only one element), then the *determinant* is defined as that matrix's element. If, however, the matrix contains multiple rows and columns, the *determinant* must be calculated.

### Determinants of 2 × 2 matrices

The determinant of the $2 \times 2$ matrix

$$A = \begin{bmatrix} a & b \\ c & d \end{bmatrix}$$

is defined as $\begin{vmatrix} a & b \\ c & d \end{vmatrix} = ad - bc.$

**Example 12:** Calculate the determinant of the matrix.

$$A = \begin{bmatrix} -2 & 3 \\ 4 & 10 \end{bmatrix}$$

Simply multiply $a_{11}$ by $a_{22}$ and subtract the product of $a_{12}$ and $a_{21}$.

$$|A| = (-2)(10) - (3)(4) = -20 - 12$$

$$|A| = -32$$

### Minors and cofactors

Two additional concepts must be defined before you can calculate the determinants of matrices larger than $2 \times 2$.

The **minor**, $M_{ij}$, of element $a_{ij}$ (which is contained within square matrix $A$) is equal to $|A_0|$, if $A_0$ is the matrix created by deleting the $i$th row and the $j$th column of $A$. According to its definition, then, if $A$ is $n \times n$, then $A_0$ must be $(n - 1) \times (n - 1)$.

The **cofactor**, $C_{ij}$, of the same element, $a_{ij}$, is defined as

$$C_{ij} = (-1)^{i+j} \cdot M_{ij}$$

In other words, the *cofactor* of $a_{ij}$ is simply the *minor* of $a_{ij}$ multiplied by either $-1$ (if $i + j$ is an odd number) or 1 (if $i + j$ is even).

Note that both minors and cofactors are defined in terms of determinants, so their values are real numbers, not matrices.

**Example 13:** If matrix $A = [a_{ij}]$ is defined as follows, calculate the minor and cofactor of $a_{23}$.

$$A = \begin{bmatrix} 3 & 2 & 6 \\ -1 & 4 & 7 \\ 5 & 1 & -2 \end{bmatrix}$$

You are asked to evaluate $M_{23}$ and $C_{23}$, the minor and cofactor of element $a_{23} = 7$. Begin by eliminating the second row and the third column of $A$ (both the row and the column containing the element 7).

$$A_0 = \begin{bmatrix} 3 & 2 \\ 5 & 1 \end{bmatrix}$$

$M_{23}$ is equal to $|A_0|$.

$$M_{23} = 3 \cdot 1 - 2 \cdot 5 = -7$$

Once you calculate the minor, it is a simple matter to evaluate the corresponding cofactor.

$$\begin{aligned} C_{23} &= (-1)^{2+3} \cdot M_{23} \\ &= (-1)^5(-7) \\ &= (-1)(-7) \\ &= 7 \end{aligned}$$

## Determinants of square matrices

To evaluate the determinant of any square matrix, follow these steps:

1. Find the row or column (if any) that contains the most 0's, to simplify your work in the later steps.
2. Multiply each element in that row or column by its cofactor.
3. Add the products generated by step (2).

This process is called "expanding a row (or column)," and it will work regardless of which row or column in the matrix you select.

**Example 14:** Calculate det($A$) by expanding one of its rows or columns.

$$A = \begin{bmatrix} 2 & -3 & 5 \\ 1 & 4 & 3 \\ -2 & -1 & 6 \end{bmatrix}$$

This example calculates det($A$) by expanding the third column.

$$\det(A) = 5 \cdot C_{13} + 3 \cdot C_{23} + 6 \cdot C_{33}$$

Calculate the cofactors, using the method of Example 13.

$$\det(A) = 5 \cdot 7 + 3 \cdot 8 + 6 \cdot 11$$
$$\det(A) = 35 + 24 + 66 = 125$$

**Example 15:** Calculate the minor of $b_{22}$ if matrix $B = [b_{ij}]$ is defined as follows.

$$B = \begin{bmatrix} 2 & 4 & 3 & 7 \\ -3 & 2 & 5 & -4 \\ 1 & -1 & -6 & 8 \\ 2 & -2 & 0 & 0 \end{bmatrix}$$

To calculate this minor, eliminate the second row and the second column and calculate the determinant of the resulting $3 \times 3$ matrix.

$$B_0 = \begin{bmatrix} 2 & 3 & 7 \\ 1 & -6 & 8 \\ 2 & 0 & 0 \end{bmatrix}$$

Notice that the determinant will be significantly easier if you expand the third row, since it contains two 0 elements.

$$M_{22} = |B_0| = 2 \cdot C_{31} + 0 \cdot C_{32} + 0 \cdot C_{33}$$
$$M_{22} = 2 \cdot 66 = 132$$

## Cramer's Rule

Consider this system of linear equations:

$$\begin{cases} a_1x + b_1y = c_1 \\ a_2x + b_2y = c_1 \end{cases}$$

You know numerous methods that can solve this system of equations, but **Cramer's Rule** presents yet another option to reach a solution.

According to *Cramer's Rule*, if you define the determinants

$$D = \begin{vmatrix} a_1 & b_1 \\ a_2 & b_2 \end{vmatrix},\ D_1 = \begin{vmatrix} c_1 & b_1 \\ c_2 & b_2 \end{vmatrix},\ \text{and } D_2 = \begin{vmatrix} a_1 & c_1 \\ a_2 & c_2 \end{vmatrix}$$

the solution to the linear system of equations is:

$$x = \frac{D_1}{D},\ y = \frac{D_2}{D}$$

assuming, of course, that $D \neq 0$.

In other words, $D$ is the determinant of the coefficient matrix. The other matrices, $D_n$, are created by replacing the $n$th column of $D$ with the column of constants.

If a system boils down to a $3 \times 3$ coefficient matrix, Cramer's Rule can be applied in a similar way, as demonstrated in Example 17. In fact, you can apply Cramer's Rule to any system whose coefficient matrix is square, but once those $D$ matrices get to be $4 \times 4$ or larger, the amount of work it takes to calculate the appropriate determinants makes Cramer's Method prohibitively complicated; you'd be better served with Gaussian or Gauss-Jordan elimination to solve such systems.

**Example 16:** Solve the system using Cramer's Rule.

$$\begin{cases} 5x + 11y = -26 \\ -7x - 3y = 24 \end{cases}$$

Begin by defining the determinants. $D$ is the determinant of the coefficient matrix, $D_1$ is the same determinant with the $x$-column replaced by the constants column (the numbers –26 and 24 on the right side of the equal signs), and $D_2$'s $y$-column gets replaced by those same constants.

$$D=\begin{vmatrix}5 & 11\\ -7 & -3\end{vmatrix},\ D_1=\begin{vmatrix}-26 & 11\\ 24 & -3\end{vmatrix},\ D_2=\begin{vmatrix}5 & -26\\ -7 & 24\end{vmatrix}$$

$$D=-15-(-77)=62$$

$$D_1=(-26)(-3)-(11)(24)=-186$$

$$D_2=(24)(5)-(-26)(-7)=-62$$

Now apply Cramer's Rule to find the solution.

$$x=\frac{D_1}{D}=\frac{-186}{62}=-3$$

$$y=\frac{D_2}{D}=\frac{-62}{62}=-1$$

**Example 17:** Solve the system using Cramer's Rule.

$$\begin{cases}4x-6y+5z=-7\\ -2x+3y-7z=8\\ -6x+15y+8z=-1\end{cases}$$

Define the determinants just as in Example 16, except this time, you'll have an additional determinant $D_3$, whose third column is replaced by the constant column.

$$D=\begin{vmatrix}4 & -6 & 5\\ -2 & 3 & -7\\ -6 & 15 & 8\end{vmatrix}$$

$$D=\begin{vmatrix}-7 & -6 & 5\\ 8 & 3 & -7\\ -1 & 15 & 8\end{vmatrix},\ D_1=\begin{vmatrix}4 & -7 & 5\\ -2 & 8 & -7\\ -6 & -1 & 8\end{vmatrix},\ D_2=\begin{vmatrix}4 & -6 & -7\\ -2 & 3 & 8\\ -6 & 15 & -1\end{vmatrix}$$

Evaluate each determinant by expanding a row or column.

$$D=108,\ D_1=54,\ D_2=72,\ D_3=-108$$

You can now solve the system with Cramer's Rule; note that the solution for $z$ exactly resembles the method used for $x$ and $y$.

$$x=\frac{D_1}{D}=\frac{54}{108}=\frac{1}{2}$$

$$y=\frac{D_2}{D}=\frac{72}{108}=\frac{2}{3}$$

$$z=\frac{D_3}{D}=\frac{-108}{108}=-1$$

# Graphs of Two-Variable Inequalities

If you change the equality sign in an equation or function to an inequality sign, such as <, ≤, >, or ≥, you drastically change the appearance of its graph. No longer is the graph a line or curve whose coordinates make the equation true. Instead, an entire region of the coordinate plane contains solutions to the system.

It stands to reason, then, that systems of inequalities will follow suit. No longer is the solution to a system the location on the graph where all the graphs in the system meet; instead, the solution is the region overlapped by all of the individual inequality graphs comprising the system.

## Single inequalities

Graphing an inequality on the coordinate plane is very similar to graphing equations. Follow these steps to sketch an accurate graph:

1. Draw the graph as it would appear if the inequality sign were replaced by an equal sign. If the inequality sign allows for the possibility of equality (only ≤ and ≥ do), then draw the graph as a solid line. If, however, the inequality sign does not make such allowances (< and > do not), draw the graph as a dotted line.
2. Notice that the graph splits the coordinate plane into different regions. Each of those regions represents a possible solution to the inequality. Choose one point that's clearly contained in each region, called a **test point**.
3. Plug each test point into the original inequality (with its inequality sign restored). If the test point makes the inequality true, then so do all points in that region, and that region is a solution to the inequality. Indicate a solution region by shading it in.

**Example 18:** Graph the inequalities.

(a) $y > \frac{1}{2}x - 3$

Begin by drawing the graph as though the ">" were "=". It's the equation of a line with slope $\frac{1}{2}$ and $y$-intercept –3. However, since the inequality sign does not allow for inequality, draw a dotted graph.

The line separates the plane into two regions, one *above* and the other *below* the line. Notice that only points in the region above the line comprise the solution; the test point (0,0) is clearly above the line, and it (like all other coordinates in that region) will make the inequality true when substituted for $(x,y)$:

$$0 > \frac{1}{2}(0) - 3$$

$$0 > -3 \checkmark$$

State your solution by shading the region above the line. See Figure 9-2.

**Figure 9-2** The shaded region is the solution region for the inequality in Example 18(a).

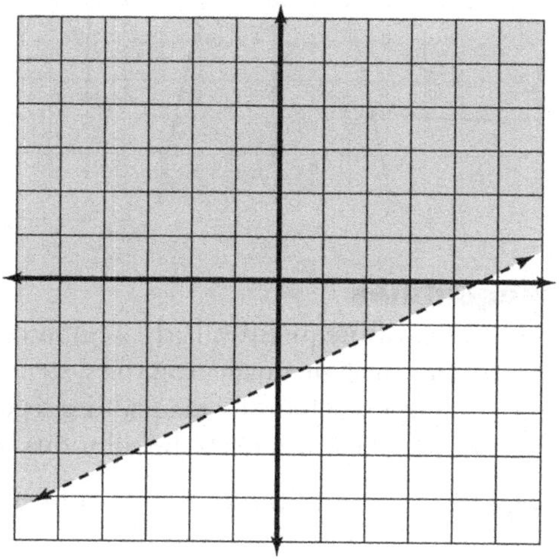

(b) $y \leq -x^2 + 2$

Begin by drawing the graph of a parabola, opening downward, with vertex (0,2); since the inequality sign is ≤, the graph should be a solid line. The graph splits the coordinate plane into two regions: one *inside* the conic and one *outside* it. If you choose test point (0,0) (which is contained inside the conic), the coordinate makes the inequality true:

$$0 \leq -(0)^2 + 2$$

$$0 \leq 2$$

Any point outside the parabola makes the inequality false, so you should shade its inside to indicate the solution (Figure 9-3).

**Figure 9-3** The solution to a conic inequality will be either the *inside* or the *outside* of its graph.

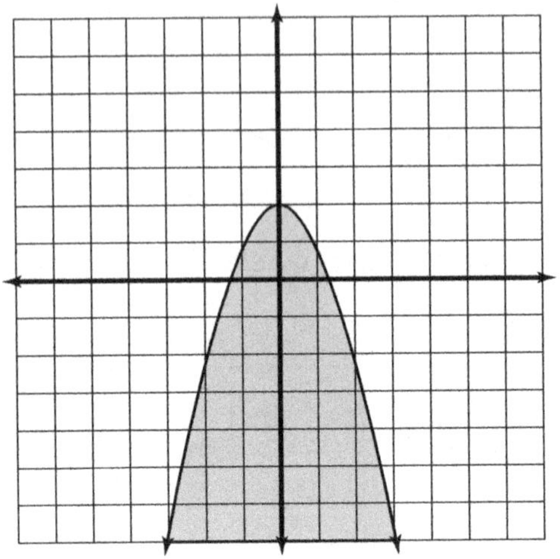

## Systems of inequalities

The solution to a system of inequalities is the region of the coordinate plane containing solutions to all inequalities in the system. The easiest way to find this region is to graph each of the inequalities separately, and then identify the portion of the plane where all the solutions overlap.

**Example 19:** Graph the solutions to the systems of inequalities.

(a) $$\begin{cases} y > \frac{1}{2}x - 3 \\ y \leq x^2 + 2 \end{cases}$$

The inequalities in this system are the same ones you graphed in Example 18 (a) and (b). Therefore, the overlapping portion of the graph will be the region inside the solid parabola and above the dotted line (Figure 9-4).

(b) $$\begin{cases} x^2 + y^2 < 25 \\ \frac{x^2}{4} + \frac{y^2}{25} \geq 1 \end{cases}$$

The first graph is a dotted circle with radius 5, centered at the origin; its solution is the inside of the circle. The second is an ellipse, centered at the origin, with vertical major axis of length 10 and horizontal minor axis of length 4; its solution is the region outside the ellipse. Therefore, the solution to the system is the region inside the circle but outside the ellipse (Figure 9-5).

**Figure 9-4** The solution to Example 19(a).

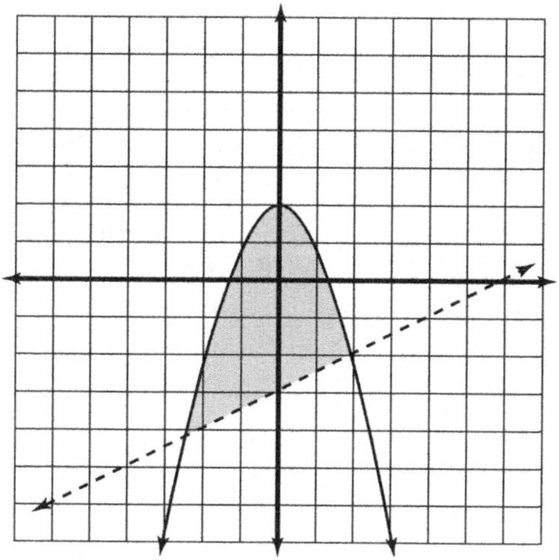

**Figure 9-5** The solution to Example 19(b).

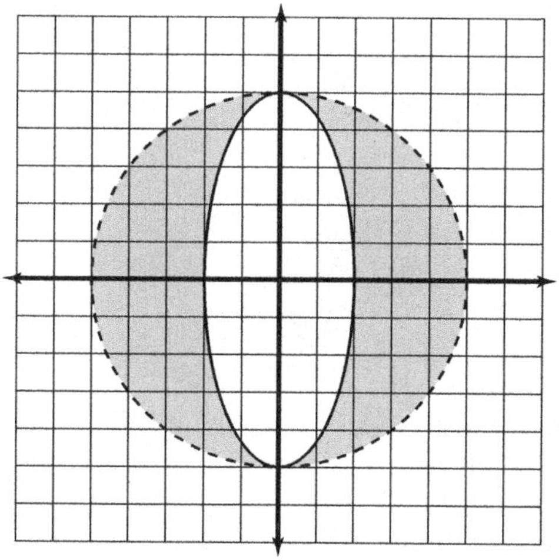

## Linear Programming

One practical application for systems of inequalities is a technique called **linear programming**. It uses a region defined by linear inequalities (called **constraints**) to find **optimal** values for a function $f(x,y)$, which is defined in terms of two variables ($x$ and $y$). (The *optimal* value for $f(x,y)$ is either the largest or smallest possible value of $f$, depending upon what is specified by the problem.)

In essence, the goal of a linear programming problem is to choose the set of coordinates (from among an infinite number of candidates) that gives the function either its largest or smallest possible value. Here are the steps you should follow when solving a linear programming problem:

1. Graph the linear inequality constraints, noting the shaded region that is the common solution defined by the system of inequalities. That region is called the set of **feasible solutions**, because the solution to the problem must come from within that region.
2. Calculate the points of intersection of the constraints. (To do so, treat each intersecting pair of constraints as a system of equations and determine the solution point.) Those points are called the **vertices** of the feasible solution set.
3. Plug each vertex into $f(x,y)$ to determine the function's value at each vertex. *The solution will always occur at one of the vertices.*
4. Identify the vertex that gives the minimum or maximum value of $f(x,y)$, as dictated by the problem.

**Example 20:** Find the maximum value of the function $f(x,y) = 2x + 3y$, if $f$ is subject to the following constraints:

$$\left.\begin{aligned} &x \geq 0 \\ &y \geq 0 \\ &2x - y \geq -2 \\ &x + 2y \leq 9 \\ &3x + y \leq 12 \end{aligned}\right\}$$

The first two constraints restrict the region of feasible solutions to the first quadrant only. Graph the remaining three inequalities to determine its other borders, and shade in the solution to the system of inequalities. There is no need to graph each line in its entirety when illustrating the region of feasible solutions; you only have to graph the line segment whose endpoints are that constraint's intersections with another contraint.

Once you have determined the feasible region, calculate its vertices (see Figure 9-6).

**Figure 9-6** The feasible region for Example 20 with its vertices labeled.

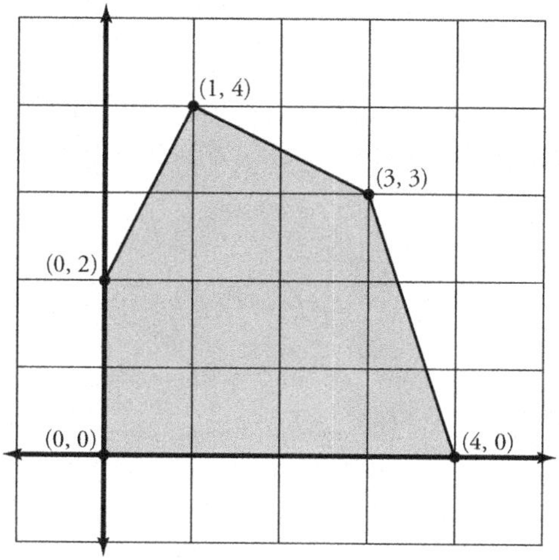


Evaluate $f(x,y)$ for each vertex.

$$f(0,0) = 2(0) + 3(0) = 0$$
$$f(0,2) = 2(0) + 3(2) = 6$$
$$f(1,4) = 2(1) + 3(4) = 14$$
$$f(3,3) = 2(3) + 3(3) = 15$$
$$f(4,0) = 2(4) + 3(0) = 8$$

The maximum value for $f$ given those constraints is 15, and it occurs when $x = 3$ and $y = 3$.

Note that the region of feasible solutions was bounded in this problem, since the region had a clearly defined perimeter and its area was not infinite. If, however, the region is infinitely large for a given problem (such that the vertices do not connect to form a closed shape) then the region can only be used to determine minimum values of $f(x,y)$, not maximum values.

## Chapter Checkout

### Q&A

**1.** Show that the solution to the system is the same if calculated using the two different methods listed.

$$\begin{cases} 3x+4y=23 \\ 5x-2y=-31 \end{cases}$$

(a) The elimination method
(b) Cramer's Rule

**2.** Given matrices $A$ and $B$ as defined below, perform the indicated operations.

$$A=\begin{bmatrix} 3 & -1 & 2 \\ 1 & 4 & -6 \\ -5 & 3 & 10 \end{bmatrix} \quad B=\begin{bmatrix} 4 & 0 & 7 \\ 2 & -9 & 6 \\ 8 & 1 & -3 \end{bmatrix}$$

(a) $2A-B$
(b) $A \cdot B$

**3.** Put $M$ in reduced row-echelon form.

$$M=\begin{bmatrix} 2 & 4 & 3 & 10 \\ 0 & 5 & 2 & 2 \\ 1 & 1 & 1 & 5 \end{bmatrix}$$

**4.** Calculate $A^{-1}$, the inverse matrix of $A$.

$$A=\begin{bmatrix} 4 & -3 \\ -2 & 1 \end{bmatrix}$$

**Answers: 1.** $x=-3, y=8$ **2. (a)** $\begin{bmatrix} 2 & -2 & -3 \\ 0 & 17 & -18 \\ -18 & 5 & 23 \end{bmatrix}$ **(b)** $\begin{bmatrix} 26 & 11 & 9 \\ -36 & -42 & 49 \\ 66 & -17 & -47 \end{bmatrix}$

**3.** $M=\begin{bmatrix} 1 & 0 & 0 & 7 \\ 0 & 1 & 0 & 2 \\ 0 & 0 & 1 & -4 \end{bmatrix}$ **4.** $A^{-1}=\begin{bmatrix} -1/2 & -3/2 \\ -1 & -2 \end{bmatrix}$

# Chapter 10
# ADDITIONAL TOPICS

## Chapter Check-In

- ❑ Applying Pascal's Triangle
- ❑ Expanding binomials via the Binomial Theorem
- ❑ Examining the terms of a sequence
- ❑ Evaluating sums and partial sums of series

The topics in this chapter deserve special mention because they either supplement the material you've learned thus far or serve to introduce elements you'll investigate more fully once you enroll in a calculus course.

The section on binomial expansions gives you great shortcuts for computing otherwise impossibly long products. The section on sequences and series is merely the briefest of introductions, just enough to whet your appetite for the more thorough approach you'll see in calculus.

## Binomial Expansion

One of the most common algebraic errors students make is improperly raising binomials to exponential powers, Remember,

$$(a + b)^2 \neq a^2 + b^2$$

Instead, you must use the FOIL technique to find the product:

$$(a + b)^2 = a^2 + 2ab + b^2$$

When a binomial is raised to an exponent larger than 2, however, the FOIL method is no longer appropriate. In such cases, you can apply one of two techniques to expand the binomial correctly.

## Pascal's Triangle

Consider the quantity $(a + b)^n$, for values of $n$ ranging between $n = 0$ and $n = 5$:

$$n = 0: \quad (a+b)^0 = 1$$

$$n = 1: \quad (a+b)^1 = a + b$$

$$n = 2: \quad (a+b)^2 = a^2 + 2ab + b^2$$

$$n = 3: \quad (a+b)^3 = a^3 + 3a^2b + 3ab^2 + b^3$$

$$n = 4: \quad (a+b)^4 = a^4 + 4a^3b + 6a^2b^2 + 4ab^3 + b^4$$

$$n = 5: \quad (a+b)^5 = a^5 + 5a^4b + 10a^3b^2 + 10a^2b^3 + 5ab^4 + b^5$$

Notice that each line of the above expansions shares these characteristics:

- Each expansion $(a + b)^n$ begins with $a^n$ and ends with $b^n$.
- The powers of $a$ in the expansion begin with $n$ and decrease by one with each consecutive term.
- The powers of $b$ in the expansion begin with 0 ($b^0 = 1$) and increase by 1 with each consecutive term.
- In each term, the exponents of $a$ and $b$ sum to $n$.

The tricky part of expanding a binomial, then, lies in calculating the appropriate coefficients, since the variable patterns are so fixed and predictable.

Notice that you can rewrite the coefficients of the expansions in a triangular pattern, called **Pascal's triangle**.

| | | | | | | | | | | | |
|---|---|---|---|---|---|---|---|---|---|---|---|
| Row 0 | | | | | | 1 | | | | | |
| Row 1 | | | | | 1 | | 1 | | | | |
| Row 2 | | | | 1 | | 2 | | 1 | | | |
| Row 3 | | | 1 | | 3 | | 3 | | 1 | | |
| Row 4 | | 1 | | 4 | | 6 | | 4 | | 1 | |
| Row 5 | 1 | | 5 | | 10 | | 10 | | 5 | | 1 |

*Pascal's triangle* has the following properties:

- Each row begins and ends with 1.
- Each row contains one more element than its row number.

- Each term in the triangle is equal to the sum of the terms to its upper-right and upper-left.
- The terms in the row $r$ are the coefficients of the binomial expansion $(a + b)^r$.

**Example 1:** Expand the binomial $(x + y)^4$.

The correct coefficients for the expansion come from row four of Pascal's triangle. (Don't forget that triangle technically begins with row zero.)

$$1 \quad 4 \quad 6 \quad 4 \quad 1$$

The powers of $x$ begin with $n = 4$ and decrease by 1, left to right.

$$1x^4 \quad 4x^3 \quad 6x^2 \quad 4x^1 \quad 1x^0$$

The powers of $y$ begin with 0 and increase by 1, left to right.

$$1x^4y^0 \quad 4x^3y^1 \quad 6x^2y^2 \quad 4x^1y^3 \quad 1x^0y^4$$

The final expansion is the sum of those terms in simplified form.

$$(x + y)^4 = x^4 + 4x^3y + 6x^2y^2 + 4xy^3 + y^4$$

**Example 2:** Expand the binomial $(2x - 3y)^3$.

The coefficients for the expansion $(a + b)^3$ are found in the third row of Pascal's triangle.

$$1 \quad 3 \quad 3 \quad 1$$

Include the $a$ and $b$ factors as you did in Example 1.

$$1a^3 + 3a^2b + 3ab^2 + 1b^3$$

For this problem, $a = 2x$ and $b = -3y$.

$$(2x)^3 + 3(2x)^2(-3y) + 3(2x)(-3y)^2 + (-3y)^3$$
$$8x^3 - 36x^2y + 54xy^2 - 27y^3$$

It is not always convenient to find the $n$th row of Pascal's triangle in order to expand binomials, especially when $n$ gets relatively large, or if you are seeking a single term rather than the entire expansion. However, before you learn another technique to expand binomials, you must first be introduced to a new concept.

## Factorials

The **factorial**, $n!$ (read "$n$ factorial"), of a natural number $n$ is defined as the product of $n$ with all of its preceding natural numbers.

$$n! = n(n-1)(n-2)...(2)(1)$$

In other words, to evaluate a factorial, multiply the given number by all of the natural numbers which are less than it:

$$6! = 6 \cdot 5 \cdot 4 \cdot 3 \cdot 2 \cdot 1 = 720$$

The only non-natural number that has a valid factorial value is 0:

$$0! = 1$$

Since 0 is not natural, its value is not derived using the usual factorial definition, but it is necessary to define this value because it surfaces in many formulas which would otherwise be undefined.

**Example 3:** Simplify the fraction.

$$\frac{10!}{6!}$$

The numerator can be rewritten so that it contains a 6!, which can then be canceled with the 6! in the denominator.

$$\frac{10 \cdot 9 \cdot 8 \cdot 7 \cdot 6!}{6!}$$

$$[10 \cdot 9 \cdot 8 \cdot 7] = 5040$$

## The Binomial Theorem

The **Binomial Theorem** states that the $(k+1)$st term of the expansion $(a+b)^n$ equals

$$\frac{n!}{(n-k)!k!}a^{n-k}b^k$$

The binomial coefficient $\frac{n!}{(n-k)!k!}$ is usually written $\binom{n}{k}$.

**Example 4:** Expand the binomial $(x+y)^4$ using the Binomial Theorem.

The variables will expand just as they did in Pascal's triangle. To calculate the coefficients, use $k$ values that begin with 0 (for the leftmost term) and increase to $n = 4$ (for the rightmost term).

$$\binom{4}{0}x^4+\binom{4}{1}x^3y+\binom{4}{2}x^2y^2+\binom{4}{3}xy^3+\binom{4}{4}y^4$$

Apply the Binomial Theorem to calculate the coefficients.

$$\binom{4}{0}=\frac{4!}{(4-0)!0!}=\frac{24}{24\cdot 1}=1$$

$$\binom{4}{1}=\frac{4!}{(4-1)!1!}=\frac{24}{6\cdot 1}=4$$

$$\binom{4}{2}=\frac{4!}{(4-2)!2!}=\frac{24}{2\cdot 2}=6$$

$$\binom{4}{3}=\frac{4!}{(4-3)!3!}=\frac{24}{1\cdot 6}=4$$

$$\binom{4}{4}=\frac{4!}{(4-4)!4!}=\frac{24}{1\cdot 24}=1$$

Substitute the coefficient values into the expansion:

$$(x+y)^4=x^4+4x^3y+6x^2y^2+4xy^3+y^4$$

Note that this is the same expansion you got in Example 1, when you used Pascal's triangle.

**Example 5:** Find the 11th term of the expansion $(x-2y)^{15}$.

Apply the Binomial Theorem with $k=10$, $n=15$, $a=x$, and $b=-2y$.

$$\frac{15!}{(15-10)!10!}x^{15-10}(-2y)^{10}$$

$$=\frac{15\cdot 14\cdot 13\cdot 12\cdot 11\cdot \cancel{10!}}{5!\cdot \cancel{10!}}x^5(-2y)^{10}$$

$$=3003x^5(1024y^{10})$$

$$=3075072x^5y^{10}$$

## Ordered Number Lists

A large portion of second-semester calculus is spent studying infinite lists of numbers and investigating things such as whether or not those infinite lists sum up to a single, finite number. For now, an introduction to these lists is sufficient.

## Sequences

A **sequence** of numbers is an ordered list (either finite or infinite) of numbers called **terms**, which are usually defined according to some rule $a_n$. The terms of a **recursive sequence** are defined based on one or more of the preceding terms in the sequence.

**Example 10:** Write the first three terms of the sequence.

$$a_n = 3^n - 2n$$

Plug in values of *n* ranging from 1 to 3 to get the corresponding terms.

$$a_1 = 3^1 - 2(1) = 1$$

$$a_2 = 3^2 - 2(2) = 5$$

$$a_3 = 3^3 - 2(3) = 21$$

**Example 11:** If $b_0 = 2$, find the next five terms of the recursive sequence.

$$b_n = b_{n-1} + n^2$$

Plug in *n* values beginning with 1 and ending with 5.

$$b_1 = b_0 + 1^2 = 2 + 1 = 3$$

$$b_2 = b_1 + 2^2 = 3 + 4 = 7$$

$$b_3 = b_2 + 3^2 = 7 + 9 = 16$$

$$b_4 = b_3 + 4^2 = 16 + 16 = 32$$

$$b_5 = b_4 + 5^2 = 32 + 25 = 57$$

## Series

A **series** is the sum of the terms of a sequence. This summation is usually written using sigma notation:

$$\sum_{n=1}^{c} a_n = a_1 + a_2 + a_3 + \ldots + a_{c-1} + a_c$$

where *n* is the **index**, *c* is the **upper summation limit**, and 1 is the **lower summation limit**.

Note that the *upper summation limit* can be ∞, in which case the series sum may or may not exist. If the upper limit is finite, then the sum of the first $c$ terms is called the **$c$th partial sum** of the series.

**Example 12:** Find the sum of the series.

$$\sum_{n=1}^{5}(-1)^n(n-1)!$$

Calculate the terms of the sequence.

$$a_1 = (-1)^1(1-1)! = -1 \cdot 1 = -1$$

$$a_2 = (-1)^2(2-1)! = 1 \cdot 1 = 1$$

$$a_3 = (-1)^3(3-1)! = -1 \cdot 2 = -2$$

$$a^4 = (-1)^4(4-1)! = 1 \cdot 6 = 6$$

$$a^5 = (-1)^5(5-1)! = -1 \cdot 24 = -24$$

The sum of the series is equal to the sum of these five terms.

$$\sum_{n=1}^{5}(-1)^n(n-1)! = -1+1-2+6-24 = -20$$

**Example 13:** Find the third partial sum of the infinite series.

$$\sum_{n=1}^{\infty}\frac{n}{n+1}$$

The third partial sum is the sum of the first three terms.

$$\sum_{n=1}^{3}\frac{n}{n+1} = \frac{1}{2}+\frac{2}{3}+\frac{3}{4} = \frac{23}{12}$$

## Chapter Checkout

### Q&A

1. Expand the binomial $(3x + y)^6$ using Pascal's Triangle.
2. Calculate the 7th term of the expansion for $(x - y)^{10}$.
3. Given the sequence $a_n = 6\left(-\frac{1}{2}\right)^n$, calculate $\sum_{n=1}^{4} a_n$.

**Answers: 1.** $729x^6 + 1458x^5y + 1215x^4y^2 + 540x^3y^3 + 135x^2y^4 + 18xy^5 + y^6$ **2.** $210x^4y^6$ **3.** $-\frac{15}{8}$

# CQR RESOURCE CENTER

CQR Resource Center offers the best resources available in print and online to help you study and review the core concepts of precalculus. You can find additional resources, plus study tips and tools to help test your knowledge, at www.cliffsnotes.com.

## Books

*CliffsQuickReview Precalculus* is one of many great books available to help you review, refresh, and relearn mathematics. If you want some additional resources for math review, check out the following publications.

**CliffsQuickReview Algebra II**, by Edward Kohn, M.S., contains lots of practice problems and examples, much like this book. In addition, second semester algebra and precalculus courses contain some overlapping topics, so you can get another instructor's take on the material. John Wiley & Sons, Inc.

**CliffsQuickReview Trigonometry**, by David A. Kay, M.S., provides an in-depth study of trigonometry. Examples abound, and the explanations are very good. John Wiley & Sons, Inc.

**Bob Miller's Calc for the Clueless: Precalc with Trigonometry**, by Robert Miller, provides yet another perspective on the study of precalculus. This book's strength lies in its non-threatening voice. McGraw Hill.

**Precalculus,** by Ron Larson and Robert P. Hostetler, is the best precalculus textbook out there. With only a few exceptions, the material is presented in a digestible and readable fashion. It contains tons and tons of practice examples with useful solutions to the odd problems. Highly recommended! Houghton Mifflin.

**The Complete Idiot's Guide to Calculus**, by W. Michael Kelley, is the best basic guide to calculus on the market. If you're a precalculus student, surely calculus looms in your future, and this book is a terrific supplement to your textbook. Alpha Books.

John Wiley & Sons also has three Web sites that you can visit to read all about the books we publish:

- www.cliffsnotes.com
- www.dummies.com
- www.wiley.com

## Internet

The Web is always a good source for free math help and tutoring. These are just a few of the sites you'll find most informative as you commute along the information superhighway:

**Mathematics Help Central** www.mathematicshelpcentral.com/lecture_notes/precalculus_algebra.htm This site provides lots of brief but useful recaps of all the major precalculus topics (as well as recaps for numerous other college math courses). Precalculus is split into two components (algebra and trigonometry), but the link above takes you straight to the algebra page.

**Precalculus Quiz Generator and Grader** www.math.ua.edu/precalc.htm Courtesy of the University of Alabama, this Web tool automatically generates and provides solutions for 15-question quizzes on the major topics of precalculus. You can even choose what types of questions that will be included.

**OJK's Precalculus Page** www.geocities.com/ojjk/ A veteran mathematics teacher created this site to provide exhaustive review for precalculus students. It contains explanations and examples for just about every topic you can think of.

**Calculus-Help.com** www.calculus-help.com Precalculus students are bound to become calculus students, and there is no site better to get practice problems or multimedia tutorials to deepen your understanding of calculus.

Don't forget to drop by www.cliffsnotes.com. We created an online Resource Center that you can use today, tomorrow, and beyond. (You can even download the majority of the *CliffsQuickReview* and *CliffsNotes* titles!)

# GLOSSARY FOR CQR PRECALCULUS

**absolute value (of a complex number)** see *modulus*.

**amplitude** the value by which the graph of a trigonometric function such as sine or cosine is stretched; the amplitude is always a positive value.

**argument** the angle measured from the positive $x$-axis to the segment joining the origin and the point representing the graph of complex number $c$.

**augmented matrix** a matrix containing more than simply coefficients; it may contain a column of solutions or even an appended identity matrix, as in the method of calculating inverse matrices.

**axis of symmetry** the line passing through the vertex of a parabola about which the graph of the parabola is symmetric.

**center (of a circle)** the point from which all points on a given circle are equidistant.

**center (of an ellipse)** the midpoint of an ellipse's major axis.

**center (of a hyperbola)** the midpoint of the transverse axis.

**circle** a set of coplanar points equidistant from a fixed point called the center.

**coefficient matrix** a matrix whose entries are the coefficients for a system of equations.

**cofactor** the value $C_{ij} = (-1)^{i+j} \cdot M_{ij}$ based upon some element $a_{ij}$ in a square matrix, where $M_{ij}$ is the minor associated with $a_{ij}$.

**cofunctions** trigonometric function pairs which differ only in the presence or absence of the prefix "co," such as sine and *co*sine.

**common logarithm** a logarithm of base 10; if a logarithm is written without an explicit base (like log $3x$), the base is understood to be 10.

**complex numbers** any number of the form $a + bi$, where $a$ and $b$ are real numbers and $i = \sqrt{-1}$. If $b = 0$, the *complex number* is also a *real number*. If, however, $a = 0$, the number is said to be *purely imaginary*.

**component form** method of writing a vector's terminal point which presupposes that its initial point is the origin.

**composition of functions** the act of plugging one function into another, usually written as $f(g(x))$ or $(f \circ g)(x)$.

**conjugate** the complex number $a \mp b$ that corresponds to any complex number $a \pm b$.

**conjugate axis** the segment perpendicular to the transverse axis at a hyperbola's center.

**constraints** linear inequalities that bound the feasible region in a linear programming problem.

**coterminal angles** angles in standard position that share the same terminal ray.

**counting numbers** the most basic set of numbers, often learned when one is first taught to count: {1, 2, 3, 4, 5, 6, ...}. They are also called the *natural numbers*.

**Cramer's Rule** a method for solving systems of equations with matrices.

**critical number** a value for which an expression is either undefined or is equal to zero.

**degree (angle measurement)** $1/360^{\text{th}}$ of a ray's full rotation around the origin.

**degree (of a polynomial)** the greatest exponent within a polynomial.

**DeMoivre's Theorem** allows you to calculate powers of complex numbers written in trigonometric form.

**dependent** describes a system of equations that has infinitely many solutions.

**Descartes' Rule of Signs** a method used to determine the number of possible positive and negative real roots of a polynomial.

**determinant** a real number that is defined for any square matrix $A$, expressed either as det $(A)$ or $|A|$.

**diagonal** the elements $a_{11}, a_{22}, a_{33}, \ldots, a_{nn}$ in the square matrix $A_{n \times n}$.

**directrix** the fixed line used to define a parabola; all points on the parabola must be the same distance from the directrix as they are from the parabola's focus.

**dot product** of two vectors, $\mathbf{v} = <a,b>$ and $\mathbf{w} = <c,d>$, is $\mathbf{v} \cdot \mathbf{w} = ac + bd$.

**eccentricity** the value $e = \frac{c}{a}$ for an ellipse which describes whether the graph tends more toward an oval or circular shape.

**ellipse** the set of coplanar points such that the sum of the distances from each point to two distinct coplanar points (called the *foci*) is constant.

**Euler's number** the irrational mathematical constant written as $e$, which has a value approximately equal to 2.71828182845904523....

**even functions** functions such that $f(-x) = -f(x)$.

**exponential function** has form $f(x) = a^x$, for some real number $a$, as long as $a > 0$.

**exponentiating** the process of raising a constant to the power of both sides of the equation in order to cancel out a logarithm. The exponentiated form of $\log_a x = c$ is $a^{\log_a x} = a^c$.

**factorial** the product of a natural number, $n$, with all its preceding natural numbers, written "$n!$".

**feasible solutions** the region for the system of inequalities which act as constraints in linear programming.

**foci (of an ellipse)** the two fixed focus points which define an ellipse.

**foci (of a hyperbola)** the two fixed focus points which define a hyperbola.

**focus (of a parabola)** the fixed point used to define a parabola.

**function** a relation in which every input results in one and only one output.

**Gaussian elimination** the process used to put a matrix in row-echelon form.

**Gauss-Jordan elimination** the process used to put a matrix in reduced row-echelon form.

**Heron's area formula** used to calculate the area of an oblique triangle given the lengths of all its sides.

**hyperbola** set of points such that the difference of the distances from each point to two distinct, fixed points (called the *foci*) is a positive constant.

**identity elements** numbers that, when applied in specific operations, do not alter the values you begin with.

**identity matrix** a square matrix which contains all 0 elements except for its diagonal, which contains only 1 elements.

**inconsistent** describes a system of equations that has no solutions.

**index** the small number outside of a radical sign.

**inverse function** the function, labeled $f^{-1}(x)$, which contains all the ordered pair of $f(x)$, with its coordinates reversed. In other words, if $f(x)$ contains $(a,b)$, then $f^{-1}(x)$ contains $(b,a)$.

**inverse matrix** the unique $n \times n$ matrix $A^{-1}$ corresponding to the $n \times n$ matrix $A$ such that $A^{-1} \cdot A$ equals the $n \times n$ identity matrix.

**irrational numbers** any number that cannot be expressed as the quotient $\frac{a}{b}$, where $a$ and $b$ are integers and $b$ is nonzero.

**leading coefficient** the coefficient in the term of a polynomial containing the variable raised to its highest power.

**Leading Coefficient Test** describes what direction (either up or down) the graph is heading at the far right and left edges of the coordinate axes.

**linear programming** technique used to optimize a function whose solution set is subject to a set of linear inequality constraints.

**logarithmic function** function of form $f(x) = \log_c x$ (read "the log base $c$ of $x$").

**magnitude** the length of a vector; the magnitude of $\mathbf{v}$ is written $\|\mathbf{v}\|$.

**major axis** the line segment (whose ends are *vertices*) which passes through the *foci* of an ellipse.

**matrix** a rectangular collection of numbers, arranged in rows and columns, surrounded by a single set of brackets on either side.

**minor** notated $M_{ij}$, and corresponding to a square matrix $A$, it is equal to the determinant of the matrix created by deleting the $i$th row and $j$th column of $A$.

**minor axis** the line segment, perpendicular to the *major axis*, which passes through the center of an ellipse and has endpoints on the ellipse.

**modulus** the distance $r = \sqrt{a^2 + b^2}$ from the origin to the point on the coordinate plane representing the graph of the complex number $c = a + bi$; also called the *absolute value* of $c$.

**natural exponential function** the exponential function with Euler's number as its base: $f(x) = e^x$.

**natural logarithm** the logarithmic function of base *e*, written "ln *x*" and read either "natural log of *x*" or "L-N of *x*."

**natural numbers** the most basic set of numbers, often learned when one is first taught to count: {1, 2, 3, 4, 5, 6, ...}. They are also called the *counting numbers.*

**oblique triangles** triangles which do not contain a right angle.

**odd functions** functions such that $f(-x) = -f(x)$.

**one-to-one** a term used to describe a function for which every output has only one corresponding input. Only one-to-one functions have inverses.

**optimal** maximum or minimum values of a function.

**order** describes how many rows and columns are in a matrix.

**orthagonal** describes two vectors which are perpendicular to one another.

**parabola** a set of coplanar points equidistant from a fixed point (the focus) and a fixed line (the directrix).

**parametric equations** two equations (usually "$x =$" and "$y =$") defined in terms of a third variable, called the parameter.

**partial sum** sum of the terms of a series whose upper summation limit is finite.

**Pascal's triangle** the triangular arrangement of the coefficients of binomial expansions; the $(n + 1)$th row of the triangle gives the coefficients for the expression $(a + b)^n$.

**period** the shortest length along the *x*-axis after which a periodic graph will repeat itself.

**periodic** describes a graph which will repeat itself infinitely after some fixed length of the *x*-axis, called the period.

**polar axis** the fixed ray in polar coordinates representing the initial side of the angle $\theta$.

**polar coordinates** coordinates in the form $(r,\theta)$, where *r* is the distance from the pole and $\theta$ is the angle from the polar axis.

**pole** the fixed point in polar coordinates from which the distance *r* to the point is measured.

**principal** the initial investment in a compound interest problem.

**quadrantal** an angle in standard position whose terminal side falls upon a coordinate axis.

**radian** measurement of an angle in standard position that, when extended to a circle of radius $r$ centered at the origin, will mark the endpoints of an arc whose length is also $r$.

**radius** the fixed distance between the center of a circle and any point on that circle.

**rational numbers** any number that can be expressed as a fraction $\frac{a}{b}$, where $a$ is an integer and $b$ is a non-zero integer.

**Rational Root Test** a method used to determine all possible rational roots for a polynomial.

**real numbers** any number which is either rational or irrational is also a *real number*, because the *real numbers* are made up by combining those two, smaller groups.

**rectangular coordinates** coordinates in the form $(x,y)$ in the Cartesian plane.

**recursive sequence** sequence whose terms are defined based on one or more preceding terms of the sequence.

**reduced row-echelon form** the form of a matrix in which the diagonal contains only 1s, all elements above and below the diagonal are 0s, and any rows containing only zeros are placed at the bottom of the matrix.

**reference angle** an acute angle that helps calculate trigonometric function values of an oblique angle.

**row-echelon form** the form of a matrix in which its diagonal contains only 1s, all elements to the left of the diagonal are 0s, and all rows made up entirely of zeros appear at the bottom of the matrix.

**scalar** term used to refer to a numeric, non-vector quantity when dealing with vectors.

**sequence** ordered list of numbers $a_1, a_2, a_3, \ldots$.

**series** the sum of the terms of a sequence.

**singular** describes a matrix that has no inverse.

**slant asymptote** a linear asymptote that is neither vertical nor horizontal.

**square matrix** a matrix that has the same number of rows and columns.

**standard form (of a vector)** describes a vector whose initial point lies on the origin.

**standard position** describes an angle whose initial side lies on the positive $x$-axis and whose vertex lies on the origin of the coordinate plane.

**synthetic division** a shortcut alternative to long division, which uses only the coefficients of the divisor and dividend; it is only applicable if the divisor is linear.

**system of equations** set of equations for which you are seeking coordinates that makes all of the equations in the set true.

**test points** points chosen based on the graph of an inequality to determine which regions of the graph (as defined by the inequality) make it true.

**transverse axis** segment passing through the foci of a hyperbola whose endpoints are the hyperbola's vertices.

**unit circle** a circle, centered at the origin with radius 1, which is used to calculate the sine and cosine values of certain angles.

**unit vector** a vector with magnitude 1.

**vector** quantity that possesses both magnitude and direction.

**vertex (of an angle)** the endpoint shared by the two rays forming an angle.

**vertex (of linear programming)** the point at which two constraints intersect.

**vertex (of a parabola)** the point at which the direction of a parabola changes.

**vertical line test** if a vertical line can be drawn through a graph, intersecting it in two or more places, then the graph cannot be that of a function.

**vertices (of an ellipse)** the endpoints of the *major axis*.

**vertices (of a hyperbola)** the endpoints of the transverse axis.

**zero matrix** a matrix of any order whose elements are all zeros.

**zero vector** written **0**, it is the vector with component form $<0,0>$; it is orthagonal to all vectors by definition, although it is not actually perpendicular to anything because its magnitude is 0.

# Index

## A



## B

# C



# D

# E

## F

# G



# H



# I

# L



# M

# N



# O



# P

# Q



# R

# S

# T

## U



## V

## W



## Y



## Z

# Notes

# Notes

# Notes

# Notes